General Ferdinand Foch, Commander-in-Chief of the Allied Armies.

THE PRINCIPLES OF WAR

By
GENERAL FERDINAND FOCH
Commander of the Allied Armies

Translated by
J. DE MORINNI
Late Major Canadian Expeditionary Force

WITH NINE LARGE MAPS

AMS PRESS
NEW YORK

Reprinted from the edition of 1918, New York
First AMS EDITION published 1970
Manufactured in the United States of America

International Standard Book Number: 0-404-02439-4

Library of Congress Number: 70-128436

AMS PRESS INC.
NEW YORK, N.Y. 10003

INTRODUCTION

THE present war has, in spite of all its novel features, shown once more that the fundamental principles of tactics remain unchanged, and that no man can understand the principles of warfare, much less be a leader of soldiers, until he has mastered these principles.

Among the officers hastily commissioned by the various countries engaged in the great struggle this fact is not always realized. There has been a tendency to concentrate exclusively on a knowledge of trench warfare and of such work as constitutes daily trench routine. Ignorance of the eternal principles of tactics in open warfare has resulted in heavy losses of life on several occasions when operations temporarily assumed the character of open warfare, and for such ignorance a heavy responsibility rests with those—usually junior officers—who considered a study of tactics unnecessary to their work in this war.

If we intend to push the Kaiser's men back from their present positions, if we even hope only to resist future large scale advances on their part similar to those of the past it is essential that everyone concerned should have the knowledge and confidence born of some study at least of the art of open warfare.

For those at home a proper understanding of military operations is impossible without some similar elementary study, and this study will be found to be amply repaid by the greater realization of what is going on "over there," of what the various moves mean and of the results attained.

As a help to all who are taking part in the struggle against autocracy and to their friends at home, these lectures of the great leader of our allied armies have been translated into English. They were given before this war began, yet General Foch seems to have foreseen it, and from his first brilliant work at the Battle of the Marne until now *he has consistently lived up to every principle* which he had laid down at that time.

It will be apparent that he considers the present period, long and costly as it has been, as but one of preparation for the decisive battle to come. His calm confidence is explained, and the reader will understand what method makes him willing, meanwhile, to sacrifice important ground in order to save the reserve which, he tells us, must be kept intact for that decisive battle.

General Foch uses historical examples in explaining his theories, but he does not indulge in highly technical language, so that this work requires of the reader no preliminary knowledge whatever of the science of war.

J. DE MORINNI, *Major,*
Late of the Canadian Expeditionary Force.

CONTENTS

CHAPTER		PAGE
	INTRODUCTION	3
I	THE TEACHING OF WAR	7
II	CHARACTERISTICS OF MODERN WARFARE	24
III	THE ECONOMY OF FORCES	48
IV	INTELLECTUAL DISCIPLINE	99
V	PROTECTION	121
VI	THE ADVANCE GUARD	147
VII	THE ADVANCE GUARD AT NACHOD	171
VIII	STRATEGIC SURPRISE	253
IX	STRATEGIC SAFETY	275
X	THE BATTLE: DECISIVE ATTACK	310
XI	THE BATTLE: HISTORICAL EXAMPLE	328
XII	MODERN BATTLE	353

THE PRINCIPLES OF WAR

I

THE TEACHING OF WAR

OVER the door of the building in which these lectures are given these words appear: "War College."

Can these words be used together: *War* and *College?*

How can we conceive the study of such action, war, which takes place on battlefields, under unforeseen conditions, in the face of danger; which takes advantage of surprise and of strength, violence, brutality, in order to create panic through this other form of action, study, which thrives on repose, on method, on thought, on reasoning?

In short, can war be taught? Does its very nature allow it to be taught?

If the teaching of war is possible, *what* does it bear on, and to what extent?

What should be the *nature* of the teaching to prepare for *action* without which everything is useless when the struggle comes? Should we use classes, books, which, once understood, allow us to proceed on a campaign with the conviction of solving difficulties as they appear and of being infallibly victorious?

Finally, to *what faculties* of your spirit should the

appeal be made in order to develop and train them, to prepare the man of action, and what predisposition is necessary on your part?

Such are the main questions to be decided in order to determine what methods to follow and to foresee possible results.

The teaching of war goes back to the most ancient times, but it is not until 1882-1883 that we find in France any efficient and practical instruction in warfare.

Where was the difficulty? Was it to be found in the nature of the matter taught, in the true *theory* of war, or was it in the *manner* of teaching this theory after it was ascertained?

The difficulty was due to both causes.

The theories prevailing among us until that time were incorrect. They truly listed the different factors which affect the result of war: superiority of morale, of knowledge, of command, of armament, of supplies, of defense, etc. They truly explained that the result depended on such factors, but they divided them all into two classes:

1. The first were the moral advantages: quality of the troops, of the command, amount of energy shown, passions displayed, etc., which cannot be figured exactly as to quantity; all of these were systematically left out of a reasoned study and of a theory which it was desired to make a scientific one of war; or rather, all of these were presumed to be equal on either side.

2. The second class comprised all material factors which influence results: armament, commissary, nature of the ground, numbers, etc., but which are far from being everything.

At the same time that moral factors were eliminated as

causes they were also eliminated as *results*. Defeat thus became the product of material factors, whereas we shall find it later to really be a purely *moral* result, the result of a state of mind, of discouragement, of fear brought on the vanquished by a combined use of moral and material factors employed simultaneously by the victor.

The theory was therefore that to be victorious one must have numbers, better armament, bases of supplies, the advantage of terrain. The armies of the Revolution, Napoleon in particular, later answered: We are not more numerous, we are not better armed, but we shall beat you because by our planning we shall have greater numbers at the decisive point; by our energy, our knowledge, our use of weapons we shall succeed in raising our morale and in breaking down yours.

In such manner these theories, believed true because founded on mathematical bases, were entirely wrong because they had not considered the most important factor of all, whether it be a question of command or of execution, the human factor with its moral, intellectual and physical aspects. They were fundamentally wrong because they tried to make of war an exact science. It was as if, in order to learn to ride and drive a horse, you were content to handle the figure of one, learning the names and positions of the different parts of its body. Who would dream of learning only thus to manage a horse, without taking into account its life, its blood, its temperament, without mounting the living animal itself?

The worst possible results came from theories of this nature. The teaching in our military schools was one bad result, as it also aimed only at the *material* side. Thus came these exclusive studies of ground, defenses, armament, organization, administration, all more or less scientific but dealing only with the physical side of war.

As to the moral side, the side which results from human action, it was neither understood nor explained. It was, at best, dimly guessed at in historical studies roughly outlined as in the historical novel of adventure, stories of marvelous achievements, unexplained and unexplainable if we do not allow for mysterious causes, directed perhaps by providence, like the wonderful genius of Napoleon—or his guiding star.

But in such a case instruction unavoidably resulted in superstition or fatalism, in the disbelief in work, in the uselessness of study, in mental laziness.

A man was gifted for war, or else he was not. It was necessary to go on the battlefield in order to find out.

The awakening came in 1870 when we found ourselves opposed to minds trained by a study of history and of particular cases. In this manner had Scharnhorst, Willisen and Clausewitz trained the Prussian mind since the beginning of the century.

In order to learn and understand war they had not been content to gaze on the tool with which it was to be waged, to assemble its material parts without taking into account the human side. In the records of history they had studied armies, troops on the move and at rest, with their needs, their feelings, their weaknesses, their powers of every nature. " Far from being an exact science, war is a frightful and passionate drama," says Jomini, and in that description is found the basic idea from which to start an efficient study.

Because of this special character of war, ignored in a scientific form of teaching, it had been said in other

THE TEACHING OF WAR

countries, particularly in France, that *War could only be learnt by war.*

I do not wish to discuss the kind of *experience* which comes from such training, the special advantage given to the mind by habit of coming to decisions in the presence of a real adversary, and especially of resisting such emotion as naturally follows a blow.

Unfortunately such a school is no school. It can be neither opened nor closed for our instruction, and it is insufficient because it would give us no preparation for the opening stages of the next war.

As a matter of fact, there is no studying on the battlefield. It is then simply a case of doing what is possible to make use of what one *knows*. And in order to make a little *possible* one must *know* much.

This explains the weakness, in 1866, of the Austrians who should have learnt from the war of 1859, when they met Prussians who had not fought since 1815. The former had waged war without learning anything thereby; the latter had learnt the art of war without fighting.

Two methods of learning war were therefore understood: the teaching on a mathematical basis of principles which neglected the human element and had shown themselves grossly insufficient, or the teaching of war by war itself. Both methods were wrong, and a new one had to be adopted which would not be based on any formula but on facts.

The facts supplied by the military history of nations were examined. In order to understand the various phenomena of war each fact was picked up and scrutinized as under a microscope; every element was considered: time, place, weather, fatigue, moral conditions, etc. Such questions were studied as had had to be solved by those engaged in past operations whether leaders of companies, of battalions or brigades or armies. The decisions made by them were criticized in the light of results obtained.

History must be the source of learning the art of war. "The more an army lacks war experience," wrote General de Peucker, "the more it needs to make use of the history of war for its instruction. Although the history of war is no substitute for actual experience it can be a foundation for such experience. In peace times it becomes *the true method of learning war and of determining the invariable principles of the art of war.*"

This education sprung from the teachings of history has resulted in a *theory* of war which can be taught and which will be taught further, and in a *doctrine* which can be practiced. In other words, there exist a certain number of invariable *principles,* of which the *application* varies according to circumstances.

Dragomirow explained the same idea in the following words:

"*Science* and *theory* are two entirely different things, for every form of art can and should have its theory, but it cannot be made into a science. . . . Nobody would think to-day of claiming that there can be a *science of war*. That would be as absurd as a science of poetry, of painting, of music. But it does not follow that there is no theory of war, just as there is a theory of the arts of peace. Such theory alone does not create the Raphaels, the Beethovens, the Shakespeares, but it endows them with a technique without which they could not attain the heights they reach.

"The *theory of the art of war* does not claim to produce Napoleons, but it teaches the properties of troops and ground. It points out the examples, the masterpieces achieved in the art of war, and in such manner it smooths the way for those who have natural military ability.

"It does not give to any man the satisfaction of thinking that he knows all there is to know when as a matter of fact he only knows a part. Recipes for creating masterpieces such as Austerlitz, Friedland, Wagram, for conducting such campaigns as that of 1799 in Switzerland or for winning battles such as that of Koeniggrätz, we cannot obtain from theory. But it does explain these models as types for study, not to be blindly imitated but rather that the pupil may absorb their spirit and obtain inspiration from them.

"If theory went wrong, it is due to the fact that very few theorists had seen war. . . ."

We find therefore a theory of war; it is made up in the first place of a number of principles:
Principle of economy of power;
Principle of freedom of action;
Principle of free disposal of power;
Principle of protection, etc.
The existence of such principles has been discussed, and later on their soundness. Napoleon, however, wrote: "The principles of war are those which have guided the great leaders whose achievements have been handed down to us by history."

Napoleon believed in principles of war. By studying the achievements of great leaders these principles are learnt. It is not surprising, therefore, that the same principles are found by us in a study of the wars of Napoleon.

We can conclude that the art of war, like every other art, has its theory, its principles, or it would not be an art.

But the teaching of war's principles does not aim at creating mere platonic knowledge. To understand the principles without knowing how to apply them would be useless, but understanding brings assurance, wise decisions, the power of action.

"When a soldier," says General de Peucker, "knows that he knows, when he feels that what he has learnt will enable him to steer easily through difficult circumstances, his character is strengthened; he acquires the ability to make wise decisions and to put them in practice efficiently.

"On the other hand, any man who realizes his ignorance or his need of advice from others is always perplexed, undecided and ready to lose all confidence.

"Strength of mind is of primary importance in a soldier, but where can energy lead if he lacks sufficient instruction to know which goal to aim for or what means can put it within reach?"

How can sound judgment and decision be trained in a school? Marshal von Moltke explains it thus:

"The teaching of military facts is especially intended to cause the pupil to use his knowledge (that is, the theory which has been taught him). But a result of this nature cannot be obtained if the instructor merely lectures and the pupil merely listens. It can be obtained, however, quite naturally when the teacher adds to his technical instruction some forms of practice in the course of which the matters taught are applied to specific cases."

Such is the method used: first learn the rules, then apply them to specific cases. It will be shown later what is meant by "specific cases."

Writing along the same lines, General de Peucker adds: "Officers must be trained constantly to act independently, in order to develop in them the power of using their theoretical knowledge in the practical questions of life. . . . To dimly realize some scientific truth does not necessarily mean that it can be found again later by reasoning. It is a long way from conception to *the precious ability of turning acquired military knowledge into the foundation of our decisions.*

"Between these two things: *scientific conception* and *art of command* there is an enormous step which the method of teaching must enable a pupil to take if it is to be an efficient method."

Thus appear both the method to be used and the goal to be reached: to pass from scientific conception on to the art of command, from a truth known and understood on to the practical application of such truth. This enormous step was successfully taken by the Prussians, as shown by the chiefs of their advance guards who in 1866, recently graduated from their schools, undertook the operations of that campaign with an assurance, and therefore an energy of execution, believed until then to belong only to men who had fought much and well.

Let us do as they did, and for that purpose make use of practical teaching embodying the application to specific cases of invariable principles gleaned from history. In this manner can experience be prepared, the art of command learnt, and finally the habit acquired of acting correctly without having to reason.

I have spoken of *specific cases* because in war there are none but specific cases; everything is individual, and there is no duplication.

The elements of a war problem, to begin with, are only seldom certain, they are never definite. Everything is in a constant state of change. These elements have therefore only a relative value instead of the absolute values used in a problem of mathematics.

Where only one company of men has been sighted at a certain hour a battalion is found when attacking shortly later.

A regiment of 3,000 rifles properly handled means, after a few days of campaigning, 2,800 rifles; with a smaller amount of care it may mean only 2,000 rifles. Variations in morale are at least equivalent. How can we then compare two regiments? They represent under a same name two bodies absolutely different. Sickness, fatigue, strain of every kind influence different units in different ways. Some of them soon cease to have any fighting value and become mere groups of starving, sick, worn-out men. The same applies to the tactical situation which varies correspondingly; the advantage of one of the opponents is not always the opposite of the advantage of the other. Suppose a convoy is to be escorted by one, attacked by the other; evidently the method of combat will not be the same for both sides. On the same ground, under the same conditions of time and space it will be necessary to proceed along different lines in either case.

The same regiment, the same battalion, will not fight in the same manner when pursuing a beaten enemy or engaging a fresh opponent, although in both cases they

THE TEACHING OF WAR 17

employ the same men, the same rifles, the same supplies.

Again, the same thing applies to two advance guard engagements: one can never be quite similar to the other as regards dispositions to be taken, because while they are both governed by similar considerations the ground varies in either case and there are differences in time and space.

Each case is therefore individual and has distinctive factors to be considered: ground, condition of the troops, tactical situation, etc., which make it a special problem. Certain considerations acquire unusual importance, others lose some.

From this lack of similarity in different problems results the impossibility of working them out from memory alone, and the only correct answer can be found in the correct use of invariable principles adapted to the special circumstances.

Invariable principles adapted to the special circumstances of every problem, does not that method take us back to the anarchy of ideas which it had been thought to replace by one general formula, a universal theory?

As a matter of fact it does not take us back to such a condition because a similarity is bound to come in the adaptation of invariable principles to different cases, as the result of a similar manner of considering the question: in a manner that is *purely objective*.

From a similar manner of considering questions will come a similar manner of understanding them, and from the similar manner of understanding comes a similar manner of action. The latter soon becomes instinctive.

In war there is but one manner of considering every question, that is the *objective manner*. War is not an art of pleasure or sport, indulged in without other reason as one might go in for painting, music, hunting or tennis, which can be taken up or stopped at will. In war everything is co-related. Every move has some reason, seeks some *object;* once that object is determined it decides the nature and importance of the means to be employed. The *object* in every case is the answer to the question which faced Verdy du Vernois as he reached the field of battle at Nachod.

Realizing the difficulties to be overcome he seeks in vain through his memory for an example or a principle which will show him what to do. No inspiration comes. "To the devil," says he, "with history and principles! after all, *what is my objective?*" And his mind is immediately made up. Such is the objective manner of handling a problem. A move is considered in relation to the objective in the widest sense of the word: WHAT IS THE OBJECTIVE?

This similar manner of considering questions and of understanding them causes a similar manner of action. But what follows is *an unrestricted application of every means to the objective sought.* The habit once formed of thus studying and acting on many specific cases, it will be instinctively and almost unconsciously that the work is done. Verdy du Vernois is an instance. "To the devil," says he, "with history and principles," yet he makes use of his knowledge of history and principles; without training along such lines, without the acquired habit of reasoning and deciding he would have been unable to face a difficult situation.

Having determined on principles and the method of applying them we cannot cease our learning. That would

THE TEACHING OF WAR

give us a dry skeleton unsuited to the nature of war and its endless variations.

To add to our studies we shall consider (also in the light of history) the higher branches of war. Teaching may here be less didactic but it will not be less profitable.

"The roads which lead to knowledge are the road of history and that of philosophy; both can and must complete one another to promote knowledge of war and to prepare power which, perfected by instruction in peace time, must finally as the art of leading troops bring in war such results as come from the wisest counsel, the strongest will" (Von Scherf).

Napoleon said: "One may teach tactics, military engineering, artillery work about as one teaches geometry. But knowledge of the higher branches of war is only acquired by experience and by a study of the history of the wars of great generals. It is not in a grammar that one learns to compose a great poem, to write a tragedy."

Which does not mean that grammar must not be learnt, but that in every art a knowledge of the principles is not synonymous of a power to create.

After having learnt our grammar and seen how it is applied we shall examine some masterpieces to see how the human mind works in the higher branches of the art, in strategy especially. You shall see how strategy is expressed and you will understand then that while it can be easily understood after it has been practiced, the practice of it is not easy.

Marshal Von Moltke explains the nature of strategy and the best means of learning to employ it:

"What is needed is, in the face of specific cases, to appreciate the situation as it is, with its unknown factors, to judge wisely of what is visible, to guess at the unknown, to come quickly to a decision, and to finally act with energy.

"One must consider two factors, the first of which is known: one's own will, the other unknown: the will of the opponent. To these must be added factors of another kind, impossible to foresee, such as the weather, sickness, railroad accidents, misunderstandings, mistakes, in short all factors of which man is neither the creator nor master, let them be called luck, fate or providence. War must not, however, be waged arbitrarily or blindly. The laws of chance show that these factors are bound to be as often favorable as unfavorable to one or the other opponent.

"*The general, therefore, who in every specific case takes, if not the best dispositions, at least efficient dispositions, has always a prospect of attaining his objective.*

"It is evident that a theoretical knowledge is not sufficient to that end; there must be a free, practical, artistic development of the qualities of mind and character, resting of course on a previous military education and guided by experience, whether it be the experience of military history or actual experience in war."

After discussing whether strategy is an art or a science, he ends:

"Strategy is a system of expedients. It is more than a science. It is knowledge applied to actual life, the development of the original guiding thought in accordance with constant changes of events, it is the art of acting under the pressure of the most difficult circumstances."

Such is the opinion of Von Moltke, known as the man who "did well whatever he did," an appreciation barely sufficient for a man who raised to the point of genius his method of serving his country.

Here is his axiom: " Take efficient dispositions," and for that purpose; " develop freely, practically, artistically the qualities of mind and character through a previous military education, either of military history or of actual experience in war."

Which is another way of saying that strategy is only the result of character and common sense; that in order to reach the field with that double faculty one must have developed it by practice, taken a military post-graduate course, studied and solved specific cases.

That is the method which we shall follow, and in the applications arising from our studies of strategy you will also see evidence of the doctrine or discipline of intellect, a similar manner of considering problems resulting from a similar manner of approaching a question: objectively; a similar manner of handling it afterwards: unrestricted application of every means to the objective sought.

Moreover, a study of history along these lines will be for us not only a means of learning but also a road to discovery, and in that manner a way of developing instruction.

The technical means of war: railroads, aeroplanes, telegraphy have increased so that: " To-day the General-in-chief can no longer direct everything. Even a genius requires a staff of helpers filled with initiative and thoroughly trained. How much more will a general not of unusual merit need to be assisted. The command of an army is too complex for a single man. At the same time, certain technical questions require special knowledge" (Von der Goltz).

And so, lacking a sufficient genius, where can we find the means of efficiently carrying through the undertaking, war with such masses of men, unless it be in a body of officers rendered efficient by method, by work, by science, guided by a similar spirit, obeying the same mental discipline, and numerous enough to handle and guide the heavy machinery of modern armies?

We must first understand truths, and therefore have an open mind, without prejudice, ready-made ideas, or theories blindly accepted merely because they rest on tradition. One standard alone, that of reason. Then we must apply these truths to specific cases, on the map at first, on the ground later, the battlefield ultimately. Let us not look for similarities, let us not appeal to our memory, it would desert us at the first cannon shot, and let us avoid all charts or formulae. We wish to reach the field with a trained power of judgment; it only needs to have us train it, to have us begin training it to-day. Let us for that purpose seek the reason of things; that will show us how to use them.

It is necessary, finally, to employ unconsciously some truths. For that purpose they must be so familiar to us as to have entered into our bones, to be a component part of ourselves.

Those are happy who are born believers, but they are not numerous. Neither is a man born learned or born muscular. Each one of us must build up his faith, his beliefs, his knowledge, his muscles. Results will not spring from any sudden revelation of light as by a stroke of lightning. We can only obtain them through a continued effort at understanding, at assimilation. Do not

THE TEACHING OF WAR

the simplest of arts make the same requirements? Who would expect to learn in a few moments, or even in a few lessons, to ride, etc.?

The work is here a constant appeal to thought: taking only preconceived dispositions, noting ideas as they come, after the study of a map determining the elements of doctrine which now seem proven, after the correction of work bringing one's ideas closer to those of the teacher.

Then only do minds stretch in accordance with the study undertaken, principles are absorbed to the extent of becoming the basis of decisions taken. You will be asked later to be the brains of an army; I say unto you to-day: *Learn to think.* In the presence of every question considered independently and by itself, ask yourselves first: *What is the objective?* That is the first step toward the state of mind to be attained; that is the direction sought, purely objective. "There is no genius who tells me suddenly and in secret what I must say or do in any circumstance unexpected by others, it is reflection, meditation" (Napoleon).

II

CHARACTERISTICS OF MODERN WARFARE

WE endeavor to study and teach War. Before beginning such study we must determine what war we speak of.

Let us therefore determine at once the general characteristics of war, and especially its objective and means in France to-day, so that we may find in such study the basis for our plans or tactics.

It is evident that if, instead of speaking in Paris, I spoke at Brussels on the subject of strategy and tactics my teaching would bear particularly on one form of war. The condition of Belgium is familiar to you: a neutrality guaranteed by Europe, which perhaps is a useless scrap of paper, but until now it has assured the existence of that little state; the close proximity of two great powers, Germany and France, from whom no great obstacle divides her, either of whom could easily conquer her if the other neighbor or friendly powers do not interfere. For the Belgian army a special theory of war would be necessary, with one very definite aim: to delay as much as possible the progress of the invading neighbor. The work would consist in seeking how the Belgian army can fill that rôle: to *avoid* any decisive engagement, to *delay* the decision of battle.

All the military plans of the nation would have to rest on the same idea: the organization, mobilization, armament, fortifications, and also the training of the troops down to the instruction of the smallest units.

If we should go from Brussels to London we should find there different conditions, different purposes. They are equally well known to you: an insular position to be maintained intact through an organization which protects it; an ambition of colonies over the seas to be maintained and developed. Different plans are needed, and a different conception of war.

Again at Madrid. There, any idea of continental conquest is laid aside for the present as the result of geographical position, the nature of the frontiers, political and financial conditions, etc. Then what does the nation expect from its army? The preservation of territorial integrity. Would not the best lesson on the art of war consist of a few pages of the country's history from 1808 to 1814?

And it is the same in Rome, in Berne. . . . So many nations, so many different conditions, so many different theories.

Do not conclude therefrom that there is no invariable theory of war, but only different contingencies. Do not either adopt the skepticism of Pascal: " Truth on one side of the Pyrenees, error on the other." Let us merely and in the first place admit the existence of a specific case in such study as we carry out. In that specific case one of the factors is naturally the geographical situation with which we must begin in order to determine the theory that will enable us to attain the goal at which we aim as a nation; geographical situation including political, financial and military condition, outline of the country, its neighbors, nature of the rights to be defended or the claims to be pushed through, all of which differentiate every country from any other.

The same is true of the date of the problem, being another factor which is of a specific nature.

For if, instead of speaking to modern Frenchmen, I had to handle the same topic eighty years earlier, on the morrow of the great lessons of the Revolution and of the Empire, without losing sight of the nature and very spirit of war I should have had to consider the state of the European continent. In presence of a Europe worn out by the struggles it had just gone through, renouncing conquests and appeals to arms, having for an ideal a Holy Alliance which aimed especially at the maintenance of kingly rights, France needed an army which would enable her if necessary to keep her rank in such a Europe of monarchial and dynastic interests, to strengthen if necessary by an exhibition of power the politics of a ministry, that is an army of convenience, or rather of convention.

Thence came special military conditions as to enlistment, training, fortifications, and also a special conception of war, sufficient to the needs of that time.

The true theory, that of absolute war which Napoleon had taught Europe, was open to concessions at such times of general exhaustion, of decreased ambitions and of reduced means. In order to conquer it is enough to be more ambitious and stronger than the opponent; it is not necessary to be very ambitious or very strong if he is still less so. That explains our successes in 1854 and 1859, both wars of purely dynastic aims.

Such a theory of war, limited in its purposes and in its means, could no longer suffice when there sprang from conservative and monarchial Europe one ambitious nation, Prussia, who wished to dominate all Germany, capable of forcing compulsory universal service, thus giving again to war a national character.

It is because we failed to recognize this total change in our neighbors and the consequences it must bring, that we who had created national warfare became its victims. To an armed nation trained to conquest, to invasion, to relentless struggle, we opposed a reduced army, recruited among the poorest and most ignorant citizens, employing the methods of the 18th century which could only have been good enough for some puny struggle of ministry.

It is because all of Europe has returned to this theory of armed nations that we must again take up to-day the *absolute* conception of war as it exists in history; and because we seek that absolute conception it is not useless to choose this or that page of our history; such or such a war, even if waged successfully, to draw therefrom the conclusions which we seek.

We shall obtain our examples and the facts on which to establish a theory from certain pages of history, from that epoch of the Revolution when the whole nation took up arms in the defense of its most cherished interests: Independence and Liberty. We shall obtain them also from that epoch of the Empire when the army born in a great crisis is taken in hand and led by the greatest military genius of all times. Of the victories which resulted, Clausewitz wrote:

"Under the energetic leadership of Bonaparte, the French, riding over the ancient methods of war, began the conquest of Europe with an unbelievable and unprecedented success. Striking everything down on their way, they have sometimes at the first blow loosened the strongest nations from their bases."

We shall see later what was meant by the ancient methods. After having thus explained the past and present, Clausewitz looks with apprehension toward the future, and adds:

"Who knows whether, in a few generations, the craze will not reappear for the old fencing and ancient methods, whether the campaigns and battles of Napoleon may not then be criticized as the actions of a barbarian?

"All efforts of military writers must caution against any such dangerous mistake. May heaven grant that our efforts shall have a healthy result on the minds of those who will in future times govern our beloved country!"

From this prayer, become a reality, has sprung the Prussian General Staff; from the abandonment of the old fencing and ancient methods; from the thorough study of the campaigns and battles of Napoleon, considered not as the actions of a barbarian but as the only means of war in the truest sense.

May we also profit thereby.

May we also begin by laying aside the out-of-date methods abandoned by the Emperor, followed by his adversaries for their doom until finally taught by experience they knew how to wage a kind of war national in its nature, a war of movement and of shock in its methods.

The out-of-date methods, as far as we are concerned, are those of a war without decisive results, with limited purpose, war of maneuvers without battles, of which the following are examples:

The type of war which Joly de Maizeroy defined as follows: "The science of war consists not only in knowing how to fight, but still more in avoiding battle, in choosing one's positions, in planning one's moves so as to reach one's goal without risk . . . let battle be given only when judged *unavoidable.*" To evade, to put off, such is the formula.

The war without battles which Massenbach held up as the highest degree of art when he said of Henry, brother of Frederick, whom he greatly admired:

"He knew how to woo fortune by bold moves; more fortunate than Cæsar at Dyrachium, more great than Condé at Rocroi, he attained, like the immortal Berwick, *victory without battle.*"

At that time, war consisted of maneuvers, of positions. And so we see him making in 1806 the most unfortunate plans; trusting in particular the safety of the Prussian army to the position at Ettersberg near Weimar, exhorting the fleeing remnants from Jena to run there, as if the mountain alone without a strong army could in some way stop the victorious flow of the French.

It is that form of war which the Marshal of Saxe himself, a man of indisputable worth but with the ideas of his time, characterized as follows: "I do not favor battles, especially at the beginning of a war. I am sure that a clever general can wage it as long as he lives without being compelled to battle."

Napoleon, penetrating into Saxony in 1806, writes to Marshal Soult:

"I desire nothing so much as a big battle."

One seeks to avoid battle all his life; the other seeks it as soon as possible.

We see again in 1814 Schwartzenberg go through Bâle, meet the obstacles of Switzerland, completely isolate his army and expose it a hundred times to the blows of even a weakened Napoleon, taking all those risks in order to have the advantage of entering France through the plateau of Langres because from the plateau of Langres flow the rivers Marne, Aube, Seine, giving a strategic key to France.

In all those plans the idea of a *result* to be obtained has disappeared. Force is replaced by appearance, the intention is taken for the fact, the threat for the blow.

" The mistake," says Von der Goltz, " consisted in putting the aim of war in the carrying out of carefully planned maneuvers, and not in the destruction of the enemy's forces. The military world has always made the same mistake when it has laid aside the straight and simple notion of the laws of war."

We can no longer follow these false methods, we who are the successors of the Revolution and of the Empire, the heirs to that art born on the field of Valmy to astonish Europe, to surprise especially the Marshal of Brunswick, a pupil of the great Frederick, and to draw from Goethe that cry: " I tell you that here, to-day, begins a new epoch in the history of the world." The wars of kings were ending; the wars of peoples were beginning.

" The French Revolution," says Clausewitz, " had given to politics and to war a different character which the great Frederick had not foreseen, just as one cannot tell on the eve of some great event in what way things will turn out.

" The French Revolution, by the strength and the energy of its principles, by the enthusiasm it created in the people, had thrown the whole weight of that people and of all its resources in the scale where had only weighed before a reduced army and the limited means of the State.

" Troubling little about political alliances in which ministries anxiously discussed of war or treaties, a discussion which weakens the State and subordinates the brutal element of the fight to the reservations of diplomacy, the French army was proudly advancing through nations and saw, to the astonishment of itself and of others, how the natural strength of a State and some simple great motive were superior to the artificial combination existing between the nations."

And elsewhere he adds: " The prodigious action of the French Revolution is certainly less due to the use of new methods of war than to an entirely changed political and administrative organization, to the character of the government, to the condition of the nation, etc. . . . If the other nations have been unable to appreciate these new conditions, if they have endeavored to meet with ordinary means a display of crushing strength, that constitutes political mistakes. . . ."

Truly a new era had begun, that of national wars which were to absorb into the struggle all the resources of the nation, which were to be aimed not at dynastic interests, not at the conquest or possession of a province, but at the defense or spread of philosophic ideas first, of principles of independence, unity, immaterial advantages of·various kinds afterwards. They were destined to bring out the interest and faculties of each soldier, to take advantage of sentiments and passions never before recognized as elements of strength.

Remember Napoleon's first proclamation: " Soldiers, you are naked, badly fed, the government owes you much, it can give you nothing. Your patience, the courage you display in the midst of these rocks are wonderful, but they will bring you no glory, no fame is yours. I wish to lead you to the most fertile plains in the world. Rich provinces, large cities shall be in your power, you will find there honor, glory and wealth.

"Soldiers of Italy, could you lack courage or perseverance?"

And from every man in those reduced battalions of famished soldiers came the answer, "Forward!"

The new kind of war has begun, the hearts of soldiers have become a new weapon.

Do you now see the contrast between the two epochs?

On one side: intensive use of human masses fired by strong feelings, absorbing every activity of society and conforming to their needs the material parts of the system, such as fortifications, supplies, use of ground, armament, encampments, etc.

On the other side, the 18th century side: regular and methodical use of these material parts which become the foundation of various systems, differing of course with time but aiming always to control the use of troops, in order to preserve the army, property of the sovereign, indifferent to the cause for which it fights but not without some professional qualities, especially as regards military spirit and tradition.

It is in that case the material side idealized for determining the actions of the combatants, of living and thinking beings. The leaders are as painters who expect their brushes to give them inspirations whereas those brushes should only interpret the inspiration supplied by an independent genius.

An inclination to return to the old fencing methods, to out-of-date principles reappears at intervals in peace-time armies which do not study history, and which thus forget what above all things gives life to war, *action* and all its consequences.

All those principles and methods are founded on material things in times of peace, on the material element which keeps all its importance in peace training as opposed to the moral element which cannot then be shown or considered.

For instance: the battle of the Alma, a duplication of which in peace maneuvers would be a victory for the Russians, a defeat for the French, the nature of the ground makes it inevitable. And the peace tactician concludes that scarpments similar to those of the Alma being insurmountable they need not be protected.

The scores made in target practice, the effects of artillery fire on the ranges, are found to make any attack impossible. Therefore one must avoid attacks, one must wait to be attacked, go back to the war of positions and of clever maneuvering, turn the enemy's flank in order to starve him, etc. Every time an improvement is made in armament the defensive is found to be compulsory.

Similar questions, on the other hand, when studied from history call for an opposite answer.

The battle of the Alma is an undeniable victory for the French. Therefore all ground is passable to the enemy unless it is defended by watchful and active troops.

All improvements in firearms add to the strength of an offensive intelligently planned because the attackers, choosing their ground, can concentrate on it so much a greater volume of fire. Moreover there is the question of morale, the spiritual superiority of the attacker over the defender, of the crusher over the crushed. The attack will need more preparation before moving its men, yet it retains the advantage even as regards the volume of fire.

There is often an inclination, even when studying history, to attribute to purely material causes the main results of any war.

"The French people," says Von der Goltz, "has always concentrated on material questions. They carefully observed the events of the war of 1866 and sought the secret of the Prussians' victory only in the superiority of their armament. By realizing in this manner only the visible side of Prussian military strength they realized also only the visible obstacles which it encountered. It was an axiom for the French army to use to its fullest extent the power of armament, and to remain strictly on the defensive. They thought that the offensive power of the German army would be broken by the defensive action of new and terrible weapons. Our opponents did along the development of this theory much more than had ever been done by any army previously, and yet victory did not reward them.

"They ruined in that way the spirit of their army, and the visible results could not fill the place of the moral strength sacrificed, of the confidence shaken. That is what chiefly weighed in the scale. *Whatever is done in an army should always aim at increasing and strengthening that moral strength.*"

In this manner, if the teaching of history is neglected, the practices of peace times bring us back slowly but inevitably to the old style of fencing through the undue importance attached to material means.

The French generals of 1870 prove that fact, as did the Prussians of 1806. On either side, as Von der Goltz says, "when the enemy threatens, the strategists study the ground, draw up imaginary plans of campaign and seek positions which they may or may not find."

Such is the brief history of our sad experience of 1870.

First, as regards positions, we find Cadenbronn, Froeschwiller, the forest of Haye, each one of which in turn is to assure the safety of the nation.

Secondly, there are imaginary plans. We shall cross the Rhine: where, when, how, by what means? that matters little. A junction with the Austrians is to be made in Bohemia. Combinations are believed valuable in themselves, without regard to time, to space, to the aim sought.

Thirdly, the idea of battle has so disappeared—and it has disappeared because it is thought unnecessary, because one hopes like the immortal Berwick to gain a victory without battle—that when troops are led to the fight success is expected from the mere disposal of these troops with regard to one another, from a perfect alignment, from some new formation or other. Battles are planned like reviews; there is no thought of the enemy or of the blows to be struck against him or of the hammer with which to strike.

These mistaken ideas will crop up often without your realizing it in your decisions; they will bring down on you our criticism when you undertake operations on the flanks or rear of the enemy whose only pretended value lies in the direction they take, when you begin threats which are not followed by attacks, when you draw up graphics and charts as though they had any intrinsic value.

All that kind of thing has no more strength than a house of cards.

A worthy opponent is not put to flight by any cleverly chosen direction. He is not nailed down without a real attack any more than a paper roof would prevent rain and cold from entering the house.

War as we study it, positive in its nature, permits only of positive answers: there is no result without cause; if you seek the result, develop the cause, employ force.

If you wish your opponent to withdraw, beat him; otherwise nothing is accomplished, and there is only one means to that end: the battle.

There is no victory without battle.

"Victory is the price of blood. One must accept the formula or not wage war. Any reason of humanity which we might advance would only expose us to be beaten by a less sentimental opponent" (Clausewitz).

Having now seen the kind of war we must not make, having compared it with the kind favored at the beginning of the 19th century, let us see what kind we should be ready for.

Let me show you first how it bears to a particularly great degree the national character of which I have spoken.

"Even the war of 1870 will be only child's play when compared to the war of to-morrow," said Bismarck.

War assumed a national character at first to conquer and guarantee the independence of peoples: the French in 1792-1793, the Spanish from 1804 to 1814, the Rus-

MODERN WARFARE

sians in 1812, the Germans in 1813, Europe in 1814, and it then showed those glorious examples of the people's passions which are known to us by the names of Valmy, Saragossa, Tarancon, Moscow, Leipzig, etc.

War afterwards became national to win the unity of races. That is the theory of the Italians and Prussians in 1866, in 1870. It is the theory by which the King of Prussia, become Emperor of Germany, will claim the German provinces of Austria.

But again we see war of a national character to win commercial advantages, favorable commercial treaties.

After having been the means chosen by peoples to win a place in the world as nations it becomes the means they choose to enrich themselves.

"Modern wars have become the business of nations. They have their interests like individuals. National selfishness is inseparable from national greatness" (Von der Goltz).

National selfishness, national greatness, such words are used jointly, and from them war is born.

The fortunes of peoples have changed like those of individuals: instead of being in land as it was or still is in absolute monarchies it has become to a great extent one of varied investments in other countries. Fortunes are represented by pieces of paper: bonds or stocks for the individuals, commercial treaties for the nations.

The means for the latter of enriching themselves, of satisfying their ambitions, is war.

The Germans by their victory of 1870-1871 acquired Alsace-Lorraine, but they also acquired the condition of an Empire, taking primary importance in Europe, providing a stronger position for each of its subjects abroad, procuring outlets to German commerce and industry because the orders of industry always come to the successful, even to those whose success is in the use of arms. They also won from France, from the protective and commercial point of view, to get the preferences of the most favored nation, that is the means of introducing cheaply in our country the output of their factories, the means of absorbing our money.

The general benefit obtained through victory is interwoven in that case with the private benefit of the individuals.

The war between China and Japan offers another example of the same kind. The Japanese, after a number of victories, signed at Simonozacki a treaty by which they received slight territorial gains but enormous commercial advantages, especially the right to enter China which insures, besides undeniable moral advantages, a great political influence throughout the Orient, the development of their commerce.

The Japanese war, inspired completely by German methods as to its preparation, its organization, its aims, gives us a small idea of the nature of modern war: a commercial undertaking by the nation, affecting the individual more closely than in the past, and therefore appealing more strongly to individual passions.

National selfishness creating a war of interest destined to satisfy the growing desires of peoples who thereby bring to the struggle an increasing flow of passions; an ever greater use and destruction of the human element and of all the country's resources. Such is the picture, and " the nations resemble men who prefer to lose their lives rather than their honor, who would rather stake their last penny than acknowledge defeat. Defeat is the ruin of all" (Von der Goltz). If the defeated side only comes to terms when it has no means left of discussion, the aim must be to destroy its means of discussion.

What means are at the disposal of such politics always more national, always more interested, always more selfish, of such war always more passionate, more violent?

" Mobilization takes to-day all intellectual and material resources of the country in order to insure success" (Von der Goltz).

They take all, as compared with previous modes of enlistment: recruiting, drawing lots, the purchase of substitutes, which even in the days of the Revolution and of the Empire left out some of the citizens.

All intellectual resources, whereas former systems allowed the escape of the more wealthy and more learned part of the nation.

But the men taken have already been taught to bear arms; they have all served a period of instruction already when they are called back to war, while the recruits of 1793 and the German Landwehrs of 1813 were all inexperienced men.

The present-day army is therefore bigger and better trained, but it is also more nervous and more easily affected. The human side of the problem which already had a greater importance than the material factor at the beginning of the 19th century must now be more important still.

Where can we study modern operations better than in the history of the French Revolution which, from the beginning, set up so high the aims of war and the means (numbers, enthusiasm, passions) employed in its service? Nowhere can better models be found than in the actions of Napoleon who made use of that wonderful military power in order to triumph by:

Taking advantage of human emotions;

Maneuvering masses of men;

Giving to operations the most crushing nature ever known.

That is why modern war draws its theories from Napoleon.

We cannot neglect either the aims of war, its purpose or the means it offers, because it is from the study of all these that we must decide on what *use* to make of such means, what tactics to employ and what objective to keep in mind.

War receives its impetus and its form from the ideas and emotions which prevail when it breaks out. That is shown by the changes which had to be undergone by the armies of Europe, fighting against Napoleon, before they achieved victory.

"Every nation," says Rustow, "gradually builds for itself a national army closely related to the country. Only then do all armies adopt the changes which the French Revolution brought in the art of war, and the spirit of the new art of command appears and takes suitable shape."

That spirit marks the end of feints, of threats, of maneuvers without battles; it embodies the only argument of blows, of battle, and the unlimited use to that end of human material, the principle of the positive purpose through the use of every ounce of strength.

"Until that spirit prevails," continues Rustow, "new

forms could be copied, but nothing could be created. The campaigns of Napoleon were imitated without results.

"But when Spain, Russia, Germany arose, they immediately found the form of war which suited each of them, and all these national armies having certain common features promptly adopted the new forms which France had brought to strategy."

Compared with that prime factor of national arisings against France, all other causes of Napoleon's defeats are negligible. If he no longer showed correct appreciation of proportions it is due in great measure to his having never considered the possibility of a national arising of his enemies, and when he suddenly saw them spring up before him he did not know how to overcome them. He could no longer neglect the ancient principle of Rome: "Never wage two wars at once."

On the other hand, it is in the ignorance of this idea that we partly find the cause of our armies' weakness on the Loire in 1870-71. The "Levée en masse," or general armed arising, ordered by Gambetta was little suited to minds trained in a school of order, method and absolute regularity.

What arguments can then be used, what methods employed in a kind of war always more national in its origin and its purposes, ever more powerful in its means, always more bitter, and which gives only secondary consideration to possession of the ground, the winning of positions or the occupation of fortified points?

As we have already seen, modern war returns to the principle of a *decision by arms*, and no other can be accepted. Instead of blaming the battles of Bonaparte as barbarous actions it recognizes in them the only efficient means; it also seeks battles in the same manner.

A modern army is faced by an opponent whose ideas of war are similar; who will be little affected by any invasion or occupation of his territory; who will only acknowledge defeat when no longer able or willing to fight, that is when his army shall have been destroyed materially or morally.

For that reason, modern war can admit of no other arguments than those which help destroy an army: the battle, the destruction by force.

"Napoleon always marched straight to his goal without in any way bothering about the strategic plan of the enemy. Knowing that everything depends on tactical results, and never doubting that he could obtain them, he has always and everywhere sought opportunities of fighting" (Clausewitz).

To seek enemy armies, nucleus of the opposing strength, in order to beat and destroy them, to follow for that purpose the direction and tactics which will lead there in quickest and surest manner, such is the lesson of modern war.

Let us therefore no longer speak of maneuvers merely intended to reach the opponent's lines of communication, to seize his stores, to enter this or that portion of his territory. None of these results is an advantage in itself; it only becomes one if it facilitates battle under favorable circumstances, if it permits the most favorable employment of forces.

Only tactical results bring advantages in war. A decision by arms, that is the only judgment that counts because it is the only one that makes a victor or a vanquished; it alone can alter the respective situation of the opponents, the one becoming master of his actions while the other continues subject to the will of his adversary. Without battle there can be no judgment, nothing is accomplished. For instance, take Valmy: Dumouriez is at Sainte-Menehould. His flank is turned, his direct communications with Paris are cut. He adopts indirect communications, and as there has been no decision by arms, no tactical result, he holds his ground. When the enemy attacks, he defends himself, and if he is not beaten it is the enemy who is beaten because he has failed at the court of battle.

There is no longer any strategy to be compared with the strategy which aims at tactical results, at victory by battle.

And since strategy has no existence in itself, since it has value only by tactics, since tactical results are everything, let us see what these tactical results are made of.

Here again "modern war is based on the ideas of Napoleon who was the first to show the importance of *preparation* and the power of *mass* multiplied by *impulsion* to shatter, in a battle sought from the beginning of the war, the moral and material resources of the adversary" (Clausewitz).

Later on, when we study the action of force, we shall conclude by reasoning to this method of understanding battle, the necessity of striking a powerful and decisive blow.

Let us, for the time being, keep to that outline drawn from history, and characterized by:

Preparation;

Mass;

Impulsion.

Preparation in modern war is more necessary, and needs to be pushed further, than in the past.

Unless that principle is accepted one is handicapped in comparison to the opponent. No longer are preparations estimated in months or weeks, but in days, hours and sometimes minutes.

"An advance of three days in the French mobilization," says Von der Goltz, "would allow the French to besiege Metz and Thionville, to sever communications with Strasburg, and to reach the Sarre before the Germans could prevent it. The latter would be obliged to concentrate back on the ground where they concentrated in 1870, that is on the Rhine."

The same applies to the point of assembly, chosen as near as possible to the frontier. It is certain that Château-Salins is 27 kilometers away from Nancy, that Nancy is occupied by a powerful French garrison, and that round Château-Salins we should see, at the very start of war, considerable mobilization of German troops.

That necessity of preparation pushed as far as possible is found in every tactical action which you will study, in order not to give the advantage to the opponent, and in order to avoid on the field mistakes which always entail great sacrifices, considering the terrible power of modern weapons.

Let us consider the tactical action itself.

Of what does it consist? We have found only one way of dealing with the opponent, which is to beat him, and therefore to knock him down. From which we gather the idea of shock made up of two things, *mass* and *impulsion*.

We have spoken of the *mass;* it absorbs for war all the physical and moral resources of the nation. The same will be true of every tactical operation, however small. The greater part of the forces, if not all, will be kept back as mass of shock.

As regards *impulsion,* one of the new ideas incorporated into war, it naturally conveys also the idea of motion. The tactics of battle must tend to movement, and the theory which seeks the hardest possible shock entails as a primary condition of strategy to bring forward all the troops. It is then by movement that troops assemble, prepare for battle. *Movement is the rule of strategy.*

But the shock could be awaited? Not indefinitely, for if it were not sought it either might not occur or occur under unfavorable conditions, and we should fail to destroy the opponent's forces, the only road to success.

We must seek shock, a new source of movements: movement to seek battle, movement to gather forces there, movement to execute it.

Such is the first law which governs the theory from which no body of troops is exempt, and which has been expressed by the following military principle: *of all mistakes, one only is disgraceful: inaction.*

This law, combined with the idea of shock, makes the leadership of troops a problem of force in time and in space, and when we remember that the movements to seek, prepare and execute battle are carried out against an opponent who is also moving, the problem becomes one of dynamics of which only one factor is known to us. We know the position and approximate power of our own forces, but we have only a vague idea of the power and position of the enemy.

The moving and unknown enemy must be therefore discovered and immobilized so that we may strike him: that necessitates a certain number of detached troops, having special missions and obliged to maneuver in order to fulfill these missions.

But any unbeaten opponent is also free to move. The meeting of forces at which we aim must therefore have protection against attacks which would otherwise prevent our marching, our assembling, our striking. That necessitates a service of protection, detached troops compelled also to maneuver.

At the same time as we seek to concentrate we seek to prevent the concentration of our opponent; at the same time as we seek to preserve our liberty of action to carry out our plan we seek to destroy his in order to strike him. Hence the need for still more detached troops.

From which we find that the primary idea of reaching the battle with the greater part of one's forces entails:

(1) The necessity of having always that greater part ready, and of maneuvering it;

(2) Also, and first of all, the need of detaching numerous bodies all destined to facilitate the work of the main body, having thereby a subordinate mission, and also compelled to maneuver. Apparently this entails dispersion instead of the concentration desired.

We shall see in the next lecture how the principle of economy of forces enables us to conciliate these contradictory conditions: *to strike with a concentrated whole* after having supplied numerous detachments.

I shall have accomplished my purpose to-day if, after

having shown you what theories to discard, I have pointed out the nature of modern war:
War ever more national;
Masses ever more considerable;
Ever greater importance of the human element;
Necessity therefore to return to the kind of leadership which seeks battle and employs maneuver to find it.
Leadership characterized by: preparation, mass, impulsion.

As regards these last characteristics, they are so deeply rooted that they will impress themselves on all the actions of war, however trifling. There will be from you no properly planned action if it does not fill all three conditions:

Preparation: that is, in your mind a plan of action founded on deep study of the objective or on the mission assigned, as well as on a thorough, careful examination of the ground, the plan being subject moreover to changes; troops disposed so as to prepare and begin the plan's execution, to picture it in a way; advance guards and flank guards in particular.

Mass: that is, a main body as strong as possible, assembled, concentrated and ready to carry out the execution of the plan.

Impulsion by which to multiply the mass, that is to throw on one objective that mass, more or less dispersed at first, reassembled later with all the means at its disposal suitably employed.

III

THE ECONOMY OF FORCES

> "The art of war consists in having always more forces than the opponent, with an army weaker than his, at the point where one attacks, or where one is attacked by him."
> —NAPOLEON.

AS has been seen previously, modern war knows only of one argument: the tactical fact, *the battle*, for which it requires *all the forces*, relying on *strategy* to bring them there, and engaging all these forces with *tactical impulsion* to arrive to shock.

That theory compels movements and maneuvering.

But at the same time as it aims at battle, it recognizes the necessity of detaching troops to
- discover the enemy;
- ascertain his strength;
- immobilize him;
- cover and protect the concentration of its own side;
- maintain the dispersion of the enemy;
- and prevent his concentration.

In opposition therefore to the theory which prescribes *concentration* arises the execution which prescribes *dispersion,* or at least much detaching of troops. Does not that show the theory to be inapplicable?

The theory appears much more inapplicable still if we remember the forces that are moved.

THE ECONOMY OF FORCES

One readily conceives how could be brought to the same battle, on the same spot, at the same time, an army like that of Turenne or of Frederick, provided with tents and stores, which advanced as a whole, whose every component part—and they were comparatively few—lived, marched and arrived easily together in sight of an enemy holding certain positions, whose immobility was an element of strength, who therefore allowed time to assemble the troops and to deploy them with proper method. But to-day, with numbers running into the millions, the armies are compelled to stretch out in order to advance, to live, to camp. They are divided into a number of columns, and each column is lengthy. The space occupied in width and depth is immense.

Two of our French army corps marching on the same road, one behind the other, with only their fighting elements, occupy approximately 40 miles. To concentrate them at the head of the column would take nearly three days.

The same applies to the extent of front which can extend conceivably over a hundred miles.

Is there any way of conducting an action which entails such forces, which is carried on over such spaces?

Our lecture of to-day seeks to show the existence of a dominating principle which enables us to carry out that theoretical play of forces, even with considerable numbers, even against an enemy who maneuvers; to spread them in time and in space, and to employ them in two different manners so as to finally apply them as desired, strategically and tactically: concentration at one point in unity of time and unity of space.

This dominating principle, known as the principle of

the economy of forces, sprang from the revolutionary period, like the difficulties which it overcame.

What is the principle of the economy of forces? A mere definition would be insufficient to explain it.

It is partly stated in the proverb that " one must not hunt two hares at one time "; one would catch neither. It is stated in the old principle of the Roman Senate: " One does not wage two wars at one time. Efforts must be concentrated." It is the rule which Frederick urged when he wrote: " One must know how to accept a loss when advisable, how to sacrifice a province (he who tries to defend everything saves nothing) and to meanwhile march with *all* one's forces against the other forces of the enemy, compel them to battle, spare no effort for their destruction, and turn then against the others."

But that is not all.

If you said that it is the art of not spending one's forces, of not dispersing one's forces, you would only say a part of the truth. You might come nearer to it if you defined it as the art of knowing how to spend, of spending to good purpose, of drawing all possible advantage from the resources at hand.

It is easier to understand what the principle is not.

"Suppose," says Rustow, " a man who divided his income into four equal parts: one for his lodging, one for his clothes, one for his food, one for his pleasures. He would always have too much for one, and especially too little for another."

That theory of fixed, invariable division will always be defeated by the theory of the *reserve on hand*.

The principle of economy of forces consists in throwing *all* one's forces at a given time on one point, in using there all one's troops, and, in order to render such a thing possible, having them always in communication among themselves instead of splitting them and of giving to each a fixed and unchangeable purpose.

The necessity of this principle became apparent from the very outbreak of the wars of the Revolution, national wars dealing with large numbers. But we must not think that it sprang suddenly and by magic from the needs of the circumstances, or that it has ever since been truly understood and followed.

As a matter of fact, when the Convention ordered the general rise to arms it brought at first only chaos in every shape and the impossibility of conducting operations, of waging war.

To create a new order of things does not mean that one can, from the start, obtain results nor even assure its existence.

Just as the political revolution, recently accomplished, might have ended after an ephemeral existence with the Directoire for instance if Napoleon had not shown the possibility of organizing by new methods a lasting power, so also without the higher minds of Hoche, Carnot, Bonaparte and other generals of the Revolution the idea of the general rise to arms, of war with unlimited resources, might have remained a chimera, a utopia destroyed by the armies and theories of the 18th century.

The first generals of the Revolution, left to themselves and although waging national war, continued to apply the war methods of the 18th century. For a long time indeed the remedy was not discovered by the average mind.

Let us remember Moreau himself who, four years after

1796, enters Germany with an army consisting by definition and rigid organization of: *a center, two wings, a reserve,* each of these parts, like each of the early armies of the Republic, having its own distinct objective.

And when he enters Germany, what do we see?

That body, made up of uninterchangeable parts, having an unvariable composition, advances, retires, takes up positions without seeking battles. Such are the operations of 1800 around Ulm, the retreat of the Black Forest, etc.

The new principles remain unknown to Moreau, as to the early generals of the Revolution, as to the French generals of the Restoration.

These new principles by which we must be inspired, imbued, whose importance and novelty we shall understand when we see with what difficulty they are understood and practiced, mean:

Instead of the lines of the 18th century, of the processional formations into wings, center and reserve of 1800, and of our own regulations until recently—

The application of the whole on one point, and for that purpose the organization of the mass into a system of attack, joined as interchangeable parts, operating separately but aiming always at the same *positive result,* at a same objective to be mastered by different means.

It is to Carnot that we first owe this manner of understanding war, of organizing it, of directing it.

All Carnot's correspondence shows him as the first to seek, in that time of revolutionary chaos, to restore order. To the spreading out and crumbling up in which will be wasted the considerable resources of France (fourteen armies in 1794) he seeks to remedy by the unity of purpose.

The numerous divisions which have been formed tend to spread out, to isolate themselves in order to live, to march, to enjoy their independence; he shows them the importance of all aiming at the same purpose.

To the block formation of former armies, impossible to reproduce and incapable of any maneuver, he seeks to substitute coördination and a similarity of efforts originating from various sources.

To assemble and to employ together troops apparently dispersed, that is the first result which he has sought and obtained.

And in the same manner at Wattignies, one of the early battles of the Revolution in which he took part, one sees the birth of the idea of attack by superior forces at one point of the line.

All that constitutes economy of forces.

He does still more, he points out how results must be sought. For instance, he writes:

"We instruct the General Officers Commanding armies operating in Germany to follow the numerous and brilliant engagements they have fought by more important battles having decisive results. Only by winning *big battles* can they completely destroy the Austrian army, and however expert that army may be in retreating from position to position, we hope that by their approach they will compel it to a general engagement whose outcome will force it to withdraw for further reorganization."

Have we not advanced considerably from the Marshal of Saxe, the good general who can wage war all his life without fighting a battle? Are we not very close to Napoleon saying: "I desire nothing so much as a big battle"; seeking, according to Clausewitz, every occasion of fighting.

Carnot writes again to Jourdan:

"What you must do is to draw the enemy into *a great and decisive battle* in his own country, on the right bank of the Rhine, and the most suitable place for you is the exact spot where he now is, namely between Dusseldorf and the Sieg or the Lahn, where he cannot fail to be destroyed if he is attacked at the proper time, and his rear pressed by General Moreau.

"Beware, my dear General, of assuming a defensive attitude, the courage of your troops would be diminished and the enemy's boldness would become extreme . . . (Something which would not have been considered before in a plan of operations, although it existed).

"It is necessary, I repeat, to deliver a big battle; to deliver it on the right bank of the Rhine, to deliver it as near as possible to Dusseldorf, to deliver it at the moment when the enemy begins to turn in order to face Moreau, to deliver it finally with *all your forces,* with your usual impetuosity, and to pursue the enemy without rest until he is entirely dispersed.

"The enemy will not fail to swing towards your left some troops to turn your flank and to stop you in your advance.

"You must have a division specially detailed to face those troops, which division shall by its strength or position disperse or hold those troops. . . ."

In what we have just read appears the idea of having battle and the idea of how to have it, to have it with all available strength by the use of masses;

THE ECONOMY OF FORCES

To devote every means to the chief objective and, in order to avoid being turned by the enemy who will not fail to try doing so, protection is necessary. But that protection must be provided by the weakest possible detachment of troops, by a minimum of forces;

To employ the greatest possible strength for the main attack, the minimum to such secondary operations as are needed to support the main attack;

Such is the principle of economy of forces put in execution.

But how is the minimum detachment of troops going to stop the forces detached by the enemy?

By its strength if sufficient, in which case it will disperse the enemy; by an impregnable position if not sufficiently strong, in which case it will be content to *hold* the enemy, which is enough for the purpose.

We thus find a new type of organized force, capable of two different actions: crushing the adversary by shock, overcoming him, and also holding him, maintaining one's position even if weak by defensive, by the use of ground according to Carnot, by the use also of maneuver later under Napoleon. That new characteristic is the *capacity of resistance* of an armed body.

That novel use of organized force, the capacity of resistance, the means of holding, together with the properties already seen of power of shock, allows the realization and employment of the principle of economy of forces.

The principle of the economy of forces having been discovered, it will enable any armed body, whatever its size, to obtain the maximum of results, to fight with the main body and to deal the opponent a blow from which he cannot recover.

Hence the words of Bonaparte to the Austrian Generals at Leoben: "There are many good generals in Europe, but they see too many things; as for me, I only see one: masses. I seek to destroy them, knowing well that the accessories will then fall of their own accord."

There are many good generals, but they endeavor to see too many things; they try to see everything, to examine everything, to defend everything, the stores, the lines of communication, this or that position. The result is attack in a number of directions, or rather a number of attacks in the offensive, it is dispersion preventing unity of command and strength of blow, it is general weakness.

Napoleon sees one thing only: masses, and he seeks to destroy them, knowing well that the accessories will then fall of their own accord.

That is the opposite principle, the destruction of the enemy's masses and therefore the use of masses to be organized.

A rule which must in future inspire all our plans and dispositions is that, in order to overcome the adversary's masses, we must assure the freedom of movement of our own. Such must be the main thought of every leader as of every subordinate.

From that aim of assuring the freedom of movement of our own troops, main purpose of the maneuver to execute, will spring all subordinate duties of detachments, advance guards, flank guards, rear guards which we include under the general heading of advance guards: auxiliary troops having always a clearly defined duty to fulfill and special tactics to observe.

But that clearly defined duty, how will the commander of the advance guard fulfill it?

With his main body.

In everything, therefore, whether it concern the direction of the mass or the leadership of a detachment, the commander of the mass, like the commander of the detachment, must:

Determine the main purpose to be served (that appears from the duty assigned);

Employ to that end the main part of his forces;

Organize the auxiliaries, supply the detachments necessary to the success of the main body;

Maintain communications between the main body and the auxiliaries, that is organize the forces in such fashion that they may eventually be used as a whole.

Advance Guards

The simplest form of protection is that supplied by Advance Guards.

Let us, for instance, suppose a body of troops at rest in the position *a, b, c,* to be guarding along D E against an enemy expected from the north, and a regiment ordered to furnish the Outposts.

The first idea will consist in extending the men uniformly along D E. If D E is 4,000 meters in length, that will allow one man per meter; that would permit resistance, but very weak resistance.

The enemy will appear, throw his advance guard along the whole front, hold the defenders along the line D E and, bringing his main body to bear against one point, L for instance, he will easily overcome what resistance he encounters there. Against one or two hundred men he will have been able to employ three thousand.

The principle of the economy of forces applied here will make use of the capacity of resistance of the troops, of their power of shock; two factors as against one.

It will say to us: Instead of holding evenly D E it is sufficient to consider along the distance D E the direc-

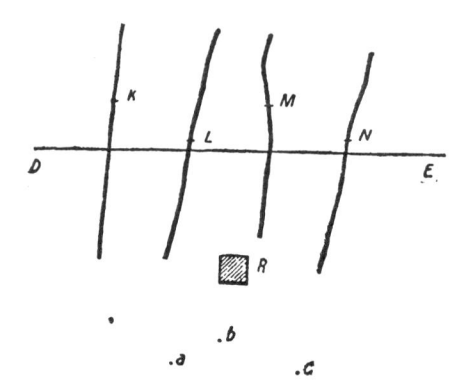

tions K, L, M, N, by which the enemy may present himself, and to hold them with troops detached at points K, L, M, N, which by their nature allow of serious resistance; locate these detachments on strong positions.

These positions being held, the whole line is held because the enemy can no longer, in view of the range of modern weapons, pass between the positions. He is therefore compelled to attack them.

Behind these detachments (Pickets) destined to resist on their positions will be located a Reserve capable of being brought to the point attacked, within the time its resistance can last, and capable of action at that point.

Because every Picket needs to be warned in time of

THE ECONOMY OF FORCES

the enemy's arrival, it will post some watchers, Sentries, which, because they need themselves to be supported and relieved, will necessitate Sentry Posts.

When the enemy appears:
The Sentries give the alarm;
The alarm reaches the Pickets and the Reserve;
The Pickets prepare to resist;
The Reserve prepares to advance.

While continuing to advance, the enemy must push home his attack in order to overcome resistance at the point which he wishes to seize. The Reserve moves to that point while it is resisting.

A regiment of 4,000 men, having four companies for outpost duty along the distance D E, for instance, will thus offer at the point of attack:

3,000 men (reserve) + 250 men (picket at point attacked) = 3,250 men as against an approximate hundred in the case of uniform distribution.

But in order to make the maneuver possible, the forces must be grouped in one system, comprising:

Pickets capable of resistance, located therefore on strong positions or on positions allowing by day good fields of fire;

Mobile Reserve, capable of maneuver within the desired space and time, having for that purpose the necessary roads; assembled; protected against the enemy's blows;

Sentries who can watch.

The same considerations apply to the defense of a stream:

To occupy on the dangerous roads strong positions (usually on the stream, because positions are found there and certain crossings only possible for the enemy) capable of a resistance which allows the reserve R to reach the point attacked, M for instance, and to employ all its forces in that direction.

The same also applies to the siege of fortified towns, which can be carried out with forces equal to the forces besieged (Metz, Paris).

A siege under these circumstances is carried out by means of a Picket line, held permanently and allowing the besieging troops to occupy, in case of attack, a First Line of Resistance organized in advance. When the attack occurs, the alarm is given by the Pickets, the First Line of Resistance is occupied, the reserves get ready. The attackers, after having fairly easily overcome the Picket Line, must, in order to continue their advance, break the First Line of Resistance; to that end they are compelled to concentrate their efforts and expose their chosen direction. The reserves and unattacked besieging troops nearest to that direction take up positions, while the resistance of the first line holds, on a Main Line of Resistance organized in advance. They offer there a resistance which gives the whole besieging army time to concentrate at some point, in the direction chosen by the attackers, and to fight there with all its resources.

Again, the same principle is followed in every attack. Then also, in order to obtain maximum results, economy of forces is used, and the troops are employed as an organized whole.

Attack could not be efficiently carried out in several directions at one time. That would divide the forces into several parts. If the enemy appears in two directions, the offensive is organized in one direction, the most

THE ECONOMY OF FORCES 61

favorable one; in the other direction he can be merely held.

In consequence, the reserves, that is the *Main Body*, are so placed and used as to push home the attack planned by the commander, and to reinforce the protecting troops if that should become necessary. As the decisive moment approaches, all these reserves move to the attack which will decide matters, and which must therefore dispose of all available forces.

Again the same principle applies to every action of the existence of troops, characterized by movement. The detachments formed for the necessary purposes which we have discussed are never more than the eyes, the fingers, the arms of one body (the Main Body) for which they work. That decides the form they should take and their close relationship with the Main Body.

They keep a relative independence, their own tactical plans, while carrying out their duties, just as my arm can protect my body, strike forward, to the right or to the left at the same time as I continue to advance.

But they must remain by the body in whose movement they have a part, from whom they draw their existence, in close enough relationship so that the body may always, in whatever direction the opponent appears, make use of all its weight. The whole system must therefore include:

Eyes turned toward the *interesting* directions;
Arms stretched towards the *dangerous* directions;
A body keeping its freedom of movement for striking in the *chosen* direction.

In consequence, a body of troops maneuvering in accordance with the principle of economy of forces does not spread itself out like the armies of the 18th century. We find it, instead, made up of a Main Body provided with such Advance Guards as circumstances require, even in actual battle as we shall see later.

These Advance Guards have every time a different composition (infantry, cavalry, artillery) suitable to the task to be fulfilled.

As to the *Capacity of Resistance* when opposed to a numerically superior enemy, it comes from:

Either the Defensive, using a strong position, and stopping the enemy unable to overcome it;

Or a *Maneuver of Retreat,* the length of which (depending on space and time) again allows the Main Body to act in accordance with established plans. In that case, the detachment does not halt the opponent, but it delays his advance.

A quick glance at the early days of the campaign of 1796 will show us very clearly the principle of economy of forces opposed to the former methods.

MONTENOTTE, DEGO, MILLESIMO

(See Map No. 1)

At the beginning of March, 1796, our forces are divided into an Army of the Alps and an Army of Italy. Conditions are equally wretched in both. Money, food, clothing, are everywhere lacking. The soldiers no longer desert; they plunder in order to exist, while their officers, equally destitute, employ the most incorrect of means. Insurrection soon appears. The army will dissolve if the most needed resources remain lacking.

The Directoire, powerless to remedy by itself an evil

THE ECONOMY OF FORCES 63

which it realizes too well, can think of no other remedy than to take the army into the rich provinces of the Peninsula, yet it is at least necessary to be able to move it. The Directoire undertakes, therefore, to have the Republic of Genoa furnish, willingly or otherwise, the most necessary help.

Schérer, overwhelmed by such a task and by so bold a plan, asks to be relieved of his command in favor of a younger and bolder chief.

Salicetti then appears as the government's representative at the Army of Italy; his efforts are at first in vain; he vainly endeavors to have Schérer alter his decision; he vainly asks the Senate of Genoa to lend a few millions to France. He is compelled to resort to threats. According to his plans, a body of French troops will occupy Voltri, and by the Bochetta it will seize the Genoese fortified place of Gavi. Schérer partly agrees with the idea, organizes an expeditionary force from the three divisions of Augereau, Laharpe, Meynier, 9,000 men altogether, and gives the command to Masséna with special instructions. The latter is to move to Savone, and to throw on March 26th at Voltri under Pijon an advance guard (3,000 men), the patrols of which will reach San Pier d'Arena, a suburb of Genoa. As for the Gavi expedition, it can wait until the arrival of Bonaparte who has just been appointed in command of the Army of Italy.

On March 26th Bonaparte reaches Nice to assume his duties. He finds there Berthier, previously requested by Schérer as Chief of Staff, attached to the new Army Commander.

Already were felt the results of the impulse given by Salicetti to all the services. There was a supply of shoes, a little money, promises of flour, of forage, of mules. It was wise to take advantage of the improvement to assume the offensive without delay.

As early as the 28th the preliminary measures are rushed through: General Headquarters is moved from Nice to Albenga; the forces are reorganized, the combatant divisions particularly being reinforced by drafts from the divisions on the coast; the cavalry is assembled by the River of Genoa; stores, munition dumps, hospitals are located; the animals necessary for the transport service are requisitioned.

The rapid execution of these measures will allow operations to begin about April 15th with 35,000 bayonets, 4,000 horses and about twenty small-caliber guns, divided as follows:

Masséna............ 18,000 men at	Savone
Augereau .... 7,000 " "	Loano
Sérurier and Rusca.... 12,000 " "	{ Garessio Bardinetto
Cavalry .... 4,800 " "	Loano

Besides:

Macquart and Garnier 7,000 men at	Tende
Divisions of the coast 9,000 " "	{ Oneille Nice Toulon

Plan of Operations

Bonaparte takes the offensive:

(1) Because the army can no longer live in the Alps, nor even on the River;

(2) Because it is the soundest tactical plan against Piedmont because of its frontier, forming a semi-circle of which the French hold the outside;

(3) Because it is in accordance with the temperament of the young General in Chief.

The Army of Italy has advanced towards the River to obtain necessary supplies; it occupies points on the coast, it keeps up relations with the Free Republic of Genoa. The offensive must start from there. What direction shall it take?

After explaining the obstacles to entering Piedmont along the heights of the Alps, Carnot in his notes of June 30th, 1794, said:

". . . If one should wish therefore to attack Piedmont, it should be through the Department of the Alpes Maritimes, taking first Oneille.

" Those reasons should decide the Committee of Public Safety to order the attack of Oneille, from where it will later be easy for us to enter Piedmont, taking Saorge in the rear and besieging Coni."

Bonaparte has a better plan. He has waged in that region the campaign of 1794 (battle of Dego). He has seen the neighborhood of Altare, Carcare, Cairo, a deep depression, 6 to 10 miles wide, giving access to the Italian valleys.

" From Vado to Ceva, first frontier post of Sardinia on the Tanaro, there is somewhat over 20 miles, without any great height.

" Snow never prevents passage there; the higher places are covered with it in winter, but in no great quantity.

" Savone, a seaport and fortified town, was in a favorable location to be used as a base. From that city to the

Madona the distance is 3 miles; a metal road joined them, and from the Madona to Carcare there are 6 miles which could be made practicable for artillery in a few days. In Carcare are found roads leading to the interior of Piedmont and Montferrat; that point was the only one at which one could enter Italy without encountering mountains " (Napoleon).

Napoleon has therefore found a passage of low height, with easy slopes and good roads from which one can easily rush at the enemy, cross the gates of Piedmont, in short assume the offensive under favorable tactical circumstances.

But there is still another advantage to marching on Carcare:

At Carcare the roads meet which

(1) By Acqui lead to Alexandria in Lombardy;

(2) By Ceva lead to Cherasco in Piedmont.

To the north of Carcare is a height of 2,000 to 2,600 feet which, over nearly 50 miles, as far as the road from Cherasco to Alexandria, prevents all communication between both provinces.

By occupying Carcare, therefore, one prevents the armies of Piedmont and Lombardy from meeting otherwise than by the road of Cherasco to Alexandria, if they intend to maneuver and fight together.

For the time being such an intention cannot be presumed. The politics of the Allies rests on individual aims, often opposed and lacking unity of purpose.

Piedmont, drawn first into the war against France, is tired of it. The people are suffering by it, and a weak government continues it merely from fear of reprisals by Austria.

THE ECONOMY OF FORCES 67

As to the latter, mistress since 1795 of Dego and Millesimo, she wishes to extend her empire as far as Savone,- on the river of Genoa, in order to obtain an outlet for the Milanese, to put up a barrier against the ambitions of Piedmont towards the Italian peninsula, and to prepare the annexation of Genoa.

The community of vues is no closer between the armies than between the governments. The conceit and ignorance of the Austrian generals intrusted with the command of the allied armies have caused much friction between both armies.

Accordingly, with the beginning of the winter of 1795-1796, the Sardinians are encamped in Piedmont, the army of Colli between Ceva, Mondovi, Cherasco. The Austrian army, beaten at Loano, has withdrawn to Lombardy and gone into winter quarters, leaving the Piedmontese alone in contact with the French.

In the spring of 1796, Beaulieu is given command of the Austrian army, Colli has kept command of the army of Sardinia.

Piedmont claims to direct operations, to have the Austrians push into Savoy. The claim is denied. The Austrian Emperor himself has refused to engage his troops beyond the Tanaro, but the Sardinian army has ceased to be under the orders of the Austrian general. From then on, the Piedmontese, in accordance with the theories of that time, will busy themselves with holding their frontiers and protecting their capital; their line of operations rests on Turin.

" They expected mediocre results from their mediocre efforts," says Clausewitz.

" It was an error to suppose that, in order to cover

Turin, it was necessary to be astride the road to that city; the armies gathered at Dego would have covered Turin, because they would have been on the flank of the road to the town" (Napoleon).

A diversity of military plans added to a diversity of political aims and an absolute ignorance of war, that is therefore the picture presented to the sight of Bonaparte.

By marching on Carcare he will not strike the center of a system of forces, but he will complete an absolute separation of interests and actions, and take advantage of it to defeat separately each of the opponents. The tactical purpose aimed at from the start: " to attack the enemy from the best direction " will be accompanied by a strategic result: " to attack a part of the forces, the Piedmontese army, finally reduced to its own means."

" By entering Italy through Savone, Cadibone, Carcare and the Bormida, one could trust to dividing the Sardinian and Austrian armies, since from there one threatened in equal measure Lombardy and Piedmont. The Piedmontese were interested in protecting Turin, the Austrians in protecting Milan " (Napoleon).

The separation being accomplished, what was the first objective to seize? The Austrian army or the Piedmontese army? Two theories are in presence: that of Carnot and that of Bonaparte.

The former, filled with his mechanical conception of forces, will push it to the degree of absurdity, neglecting to consider a special geographical and political condition which will not escape Bonaparte. So much is it true that in war there is no universal system; the most assured of principles need to be applied according to circumstances.

THE ECONOMY OF FORCES

The theory of Carnot is: "After having mastered Ceva and approached the left of the Italian Army of Coni . . . the General in Chief will direct his forces against the Milanese and especially against the Austrians. . . ; he will not forget that the Austrians are the principal enemy to defeat. . . ."

To march against the main army, to defeat it, was to end the war according to Carnot because there was the center of gravity of the resistance to be overcome.

That theory would have been true if both armies had formed a system, that is one whole of forces joined to operate together, whereas they represented two distinct unities, with different interests, two separate masses, each having its own center of gravity and its own attractions. It meant two opponents to defeat separately, two questions to handle differently; one could not end the war by striking one of the armies, even the stronger one.

On the other hand, if one chose the stronger one as first objective, one could not neglect the weaker one which maneuvered independently. It was necessary to see to what extent it might oppose the action undertaken. The geographical position of Piedmont and the condition of the Sardinian army appear here distinctly.

As Bonaparte points out, the French, in order to continue to act against the Austrians with such a precarious line of communications as that of the River, destined to stretch into Lombardy and Venetia, had to protect themselves efficiently against the Sardinians; they must employ, therefore, for the protection of that line some forces of which the absence would, in their condition of numerical weakness, be doubly felt on the principal theater of operations, preventing there any decisive result.

The necessity was therefore imperative of settling first the Sardinian question. One could only attack the Austrians after having beaten and destroyed the Sardinian army; the road to the invasion of Lombardy and Venetia ran through Piedmont. The case was a peculiar one, as every case is.

Such was, in a general way, the carefully thought out plan of Bonaparte; he will be capable of every effort to carry it through: he is twenty-seven years old, has nothing to lose and all to gain, and he is capable of every independence for he already considers as his equals the masters of France. He will dare everything.

Opposite him, Beaulieu has just assumed command of the Austrian army (he no longer commands the Sardinians); he is seventy-two years old, has a position and reputation which must not be compromised. What will he seek? First, to risk nothing, neither his reputation nor the army, nor the interests of the monarchy, though in such a way no benefit should come to any of these.

Beaulieu's plans, like his ability, are inferior to Napoleon's. He intends to assume the offensive, however, but in order to push the French back from the River; seize the Alpes Maritimes; reduce the frontage to be defended; join the English; continue the war later on the heights, and if necessary worry the French in Provence.

That idea of war differs absolutely from the novel idea started by Carnot, " the pursuit of the enemy to his complete destruction." Beaulieu aims only at partial results. His preparation and execution will, in the same manner,

THE ECONOMY OF FORCES

entail only a limited and partial use of the means at his disposal.

For instance, Beaulieu learns at Alexandria, towards the end of March the threats made by the French against Genoa, the news of an expedition against Gavi, the occupation of Voltri. He decides to attack. Without assembling his forces, disposing only of half his troops, he proposes to attack; will he, at least, bring to the attack half of his available forces? That is what we shall see later.

Anyway, he considers the French army incapable of serious resistance at present. He will consequently be able, without endangering himself, to:

Strike the right wing which has ventured toward Genoa;

Protect Genoa, whose weakness one fears;

Make connections with the English Admiral;

Avoid engaging the main French body.

By that plan again the old general is behind the times, for, in modern warfare, to set fire at one point is to start a general conflagration. When an advance guard is attacked, the army hastens to its help.

In accordance with his scheme, he sends 10 battalions, on March 31st, toward Novi, Pozzolo, Formigaro, and thence to the Bochetta (April 2nd). He also sends 11 battalions under Argenteau toward Sassello; they are spread out in a number of camps.

From the 2nd to the 9th he makes no move; he completes the preparation of his plan, confers with the English about preparing a trap for the Laharpe division and destroying it by a treble attack.

Finally, about the 5th or 6th, he settles his plans and gives his orders.

A central body under Argenteau, made up of 4 Piedmontese battalions and 12 Austrian battalions of which part will gather at Acqui, will march by Montenotte against Savone, in order to seize at that point the road of the Corniche.

A left body (10 battalions) assembled at Novi will proceed toward Voltri and attack it. The English fleet will assist in the operation to the best of its ability, by the fire of its guns or by landing men. The army of Sardinia is no longer under the orders of the Austrian general, but good relations are maintained with General Colli who commands it. Communication is less stiff but more rare than in the past. The two Generals should have conferred before acting, but since the end of March, when Colli has carried out against Sérurier some important reconnaissances, communication has been interrupted between both Generals in Chief; an appointment has been made for April 14th; by that time, events will have progressed. The Sardinian army ignores what is going on; it continues to occupy Ceva and to guard the Tanaro. It has only received from Beaulieu a request for 4 battalions to be sent to Dego under the orders of Argenteau.

While his opponents have this advantage of earlier offensive, what is Bonaparte doing?

To the first operations of the enemy, stirred by the unfortunate attempt against Genoa and the expedition to Voltri, he will answer by parades which allow him to carry out his plan.

The Sardinians are worrying Sérurier's division by raids. He reinforces that division by the Rusca brigade pushed to Bardinetto.

THE ECONOMY OF FORCES 73

The system of outposts of Schérer is carefully inspected and improved.

Masséna disposes in Savone of the 3,000 men of the Laharpe division and Ménard brigade, but in front of him, besides Cervoni at Voltri, extends a line of outposts strongly established on fortified points: these are the outposts of Stella, Montenegino, Altare, Monte Baraccone, San Giacomo, Madona della Neve, Medogno. The last four are held by the Joubert and Dommartin brigades which connect at Tanaro with the Rusca brigade.

Further back are some strong supports, near Stella, at the Madone of Savone, at Cadibone, at Quiliano.

Strong reconnaissances are in contact with the enemy as soon as he maneuvers.

Finally, Bonaparte has issued his orders himself; they entail, particularly, that Sérurier and Rusca shall remain strictly on the defensive, in order not to draw the enemy's attention towards Bardinetto, through which it is intended to reach Millesimo.

They provide for Cervoni, who has taken Pijon's place at the head of the Voltri detachment, a special mission. The Austrians seem inclined to action towards Genoa (concentration of their forces and General Headquarters at Novi). Cervoni is to draw them in that direction and hold them there; he is given for that purpose an additional half-brigade; he will carry out some reconnaissances toward San Pier d'Arena but is not to let himself be crushed by Beaulieu, withdrawing in time on Varazze, where he will be protected by the troops of Stella.

Here is a good example of an army in *defensive strategy,* in ambush so to speak, but capable of assuming the offensive. For it is protected in every direction at a distance and by forces which allow it, if attacked, to concentrate as a whole in safety to meet the attack at the point where it occurs, or to safely avoid such attack if that be the commander's intention.

Should it wish to assume the offensive, in any direction at all, the facilities for concentration are the same, and moreover the important outlets are held.

Thanks to that organization as a body (the main body) and arms (the advance guards), and to the zone of maneuver resulting therefrom, it is always able to *strike with all its mass at one point.*

We shall see, on the other hand, the three corps of Beaulieu working independently of one another, in three directions which do not connect, with three separate geographical objectives; with absolutely rigid plans moreover.

Bonaparte explains the condition clearly when he writes:

"Beaulieu was dividing his forces, inasmuch as any communication could not be maintained between his center and his left except behind the mountains, while the French army, on the other hand, was able to concentrate within a few hours and throw itself as a mass against one or the other of the enemy's corps; and, one of them being defeated, the other was inevitably compelled to retire."

After having thus guarded against possible enemy attack, Bonaparte, angered at first by the expedition to Voltri which might have precipitated events before he was ready, seeks to utilize that expedition as soon as his degree of preparedness allows it, even if he should to that end partly change his original plan without losing the

THE ECONOMY OF FORCES

main idea. By means of reconnaissances he keeps informed of the situation at Cairo, which remains unoccupied between Colli and Argenteau.

On April 9th, the Austrians resume their advance, still without advising Colli. Reconnaissances are carried out against our outposts, in particular against Cervoni. Bonaparte hastens from Albenga to Savone, coming nearer to the point where the enemy strikes.

BATTLES OF VOLTRI

Cervoni holds his positions on the 9th, with his right at Pegli, his center at Mount Pascin and at Pra di Melle, his left at Bric Germano. He has supports at Arenzano and Varazze.

On the 10th, the Austrians appear in two columns of approximately equal strength:

(1) *By Pontedecimo* comes Pittony with the artillery and cavalry (4,200 men) guarded to the right by flankers who cross San Carlo and Sant Alberto. He is stopped at Pegli by Lannes with the grenadiers of the 70th and 99th half-brigades; he shells them, and the battle remains undecided till nightfall.

(2) *By Masone* Sebottendorf with about 3,200 men attacks vigorously between Inferno and Acqua Sancta, carries the position of Pra di Melle, surrounds near Melle four companies of the 70th half-brigade which succeed, however, but not without losses, in reaching the Mount of Capucins to the north of Voltri.

Cervoni has reached the heights of Germasso; from there he surveys the movements which might threaten, by the heights of Del Dente and Reisa, his means of retreat. As the enemy makes no attempt in that direction, he leaves his troops to their fate, allows them to withdraw in accordance with local opportunities, and is content to rally them at seven o'clock that evening around Voltri. Then, after having lit some big fires, he retires, according to orders at ten o'clock on Arenzano which he has already had occupied. His movement is protected by three companies of grenadiers which fight for some time in the convent of the Capucins at Voltri and abandon the locality at midnight in order to retire on Arenzano, where they are gathered up by a rear guard (a battalion of the 99th). The retreat continues in the same manner on Varazze, which Cervoni reaches without being pursued, having lost only between 100 and 200 men, and from where he connects with the outpost of Stella.

His tactics clearly show what tactics should be employed by a retreating advance guard: protection of the line of retreat; occupation in good time of its most important points; watchfulness over the enemy's movements which might endanger it; lack of reinforcements for the troops under fire which one intends to withdraw; their successive withdrawal under the protection of covering troops; final withdrawal of the main body, without the enemy's knowledge if possible, and under protection of a rear guard which is picked up later.

Beaulieu enters Voltri at midnight. He has 7,000 to 8,000 men. He speaks of an offensive, but as a matter of fact he does nothing more; on the morning of the 11th, after an interview with Lord Nelson, who had come during the night of the 10th-11th, he keeps his troops on their positions. The result at which he aimed: to cover

Genoa, to connect with the English, seems attained; why should he act?

On the other hand, the tactical results which he expected to accomplish: to crush Cervoni with three columns and the English fleet, is not obtained because his enemy, instead of remaining immobile, held to the position, is retiring, maneuvering. But he can in that way become a danger to the troops of Argenteau from whom one has no news. Anxiety seizes Beaulieu; he perceives that he has perhaps only compromised his army; he drives off at two o'clock that afternoon in a carriage from Voltri to Novi and Acqui, where it is undoubtedly necessary to concentrate his forces. For Argenteau immediate assistance must be prepared; Beaulieu sends Wukassowitch with three battalions to support him without delay.

Thus began the downfall of the theory of partial means and partial results, of the conquest of geographical objectives. We shall see the opposite application of the theory of *absolute* war.

BATTLE OF MONTENEGINO

What is happening on the side of Argenteau, that his action is not felt on the road of the Corniche?

Argenteau receives only on the 9th the orders for his action on the 10th, instructing him to move toward Montenotte, to seize the heights held by the French posts, and to thus establish communications *perhaps* with the left body.

Under these circumstances he is unable to attack on the 10th. He employs that day in gathering the forces which he thinks he can use; and it is only on the 11th, while Beaulieu leaves Voltri, that he attacks himself at Montenotte.

On that day, early in the morning, he has started three columns on the move:

Left column (4 battalions under Lieutenant-Colonel Lezeni) leaving Sassello, through the col of Giovi, toward Stella; after encountering resistance first at San Giustino, it is definitely stopped at Stella by the 14th half-brigade, come from Savone, which holds that point with a recently-arrived detachment of the Cervoni brigade.

Center column (3 battalions) formed in Paretto and Moglio, arriving under Argenteau at Pontinvrea where it is subdivided, a detachment of 2 companies going up the right bank of the Erro while the remainder continues to Garbazzo (11 hours), where it joins the

Right column (3 companies and 2 battalions), arrived from Cairo and Dego by the heights, under Rukavina.

On that same morning, the brigade commander Rampon, sent from the Madone of Savone on a reconnaissance toward Montenotte with the 2nd battalion of the 21st half-brigade and three companies of the 1st Light Infantry, had occupied Ca Meige Dett' Amore, Cascinassa and Crocetta. About 10 o'clock the first of these posts was attacked and carried by Rukavina. Rampon retires on the Bric Castlas, which he soon abandons on the entrance of Argenteau into Cascinassa. He then withdraws on Ca di Ferro, where he holds long enough to rally his last post of the Crocetta; then, all his troops being assembled (9,000 men), he resists on Monte Pra; finally, about one o'clock in the afternoon, he is pushed back on his position at Montenegino.

Montenegino dominates the crest which extends from Mount San Giorgio to Mount Cucco. It is strengthened by a redoubt with a spear-head covering the northern slopes towards the col between Monte Pra and Montenegino, with a small redoubt to the south, 80 feet below the first one facing the western slope.

Montenegino was held by an outpost of about 600 men (2 battalions of the 1st Light Infantry, the 3rd occupying the Doria Palace).

Rampon, disposing then of 1,500 men, including those he brings, soon loses the spear-head, but he stops all evening the attacks of the Austrians (about 4,000 men). Argenteau gives up the attack until his artillery arrives; he bivouacs on Monte Pra, opposite the enemy, asking for his protection:

From Lezeni, 1 battalion which will occupy the Bric Sportiole; 2 companies to the Bric Mindo towards Altare;

From Squanello, 2 divisions and 1 battalion to the Bric Castlas.

At five o'clock, Rampon sends a note to Bonaparte, declaring himself capable of throwing the enemy back on Montenotte if he were reinforced by 1 or 2 battalions and 2 or 3 guns. The plan of Bonaparte is more important.

He has hastened during the day to the sanctuary of the Madona (1,300 feet below Montenegino) to join Laharpe. He has learnt there the incidents of the fight; he has also received reports from the outposts and the spies;

the Piedmontese have not stirred, their main body is still between Ceva and Mondovi, Carcare and Cairo are unoccupied, 2,000 men are dispersed between Dego, Millesimo and Montezemolo.

The Austrians have attacked at Voltri with 7,000 to 8,000 men, at Stella with 3,000 to 4,000 men, at Montenegino with about 4,000 men; three columns are therefore marching on Savone.

The space at his disposal no longer allows him to defer the beginning of operations. He must act at once, lest he be enveloped by Beaulieu. But he has every opportunity of bringing, on the 12th, to the battlefield which he will choose, to the direction he may name, almost all of his forces, while Beaulieu, with the greatest diligence, is unable to collect half his, because of their dispersion shown above.

During the day he has sent to all his troops a preliminary order to assume a state of readiness.

He starts again for Savone with Laharpe, calls there Masséna, explains to them verbally his intentions; Berthier communicates them immediately to the other divisions without waiting for the arrival of General Headquarters which, summoned from Albenga during the morning, can only be prepared to operate in Savone about midnight. (That mode of procedure will be usual for the Emperor.)

In accordance with these orders:

Rampon, reinforced to 2,300 men and 2 guns, spends the night at Montenegino; further back, Laharpe, after having dropped a few companies at Savone and 1 battalion at the Madone de Savone whence he will connec* with Masséna by occupying Monte Occulto, comes at

THE ECONOMY OF FORCES

midnight to the Doria Palace with about 7,000 men, namely:

The 14th half-brigade arrived late from Stella where it has merely left a detachment;

The Cervoni brigade (70th and 99th half-brigades) which Bonaparte has reviewed during the evening in Savone.

These troops will attack Argenteau in the early morning, on the front and on the flank.

Masséna, with the Ménard brigade (from the broken-up Meynier division) comes at midnight to the Plan del Melo; as soon as Laharpe advances he will attack at daybreak, towards Montenotte, in order to cut off Argenteau and overthrow all reinforcements which might reach him.

The front line troops are to bivouac about midnight to the south of Altare:

Augereau at Mallare with 6,000 men;

Dommartin with 3,000 men at Monteffredo;
Joubert with 2,000 men at Altare;
where they will await further orders.

The reserve of artillery following.

Augereau "will leave Mallare at 5 A.M. and proceed to Cairo. He will throw out a Flank Guard to his left, and will occupy the chapel of Sainte-Julie between Carcare and Cairo. If the enemy should be there, he will attack and expel him.

"Arrived past Cairo, he will occupy the mountains on the left and send patrols to Rochetta, halfway to Dego, where he will receive further orders.

"On the way, he will attack and defeat the enemy if

he meets him, and will send news of his arrival to Altare, where General Headquarters will be located."

Sérurier is urged to do much reconnaissance work.

In that way, early on the 12th, Bonaparte expects to have:

At Montenegino	Rampon	} 9,300 men
At Doria Palace	Laharpe	
At Plan del Melo	Masséna	3,500 "
On the way to Carcare	Augereau	6,000 "
At Monteffredo	Dommartin	3,000 "
At Altare	Joubert	2,000 "

23,800 men

besides the artillery reserves.

The distances between these places do not exceed five miles. Communications are assured for the transmission of orders and information.

It is with 12,000 to 13,000 men, therefore, that Laharpe and Masséna will attack the 3,000 or 4,000 men whom Argenteau has displayed on the 11th. But if, by some prodigy of activity, Beaulieu has succeeded in raising that figure to 12,000 or 15,000 men, he can still be beaten during the day with 24,000 men. The whole army (all protecting units have been recalled that were not indispensable) can maneuver toward Montenotte against reduced forces. Tactical results are undoubtedly assured.

But at the same time as he plans this tactical success, Bonaparte intends to push as much as possible his strategic plan: to separate the Austrians from the Sardinians by occupying Cairo, which was still free on the 11th. For that purpose, he sends Augereau there as early as

THE ECONOMY OF FORCES

the morning of the 12th. Any combination, indeed, can be parried if time is allowed the enemy, if he is not beaten by surprise. As Gneisenau wrote: " Strategy is the art of utilizing time and space. I am more careful of one than of the other. Space I can always find again; wasted time, never."

But if the enemy should have been warned, and have occupied Cairo at the last minute, he will be thrown out before he can be reinforced there. Augereau will present himself with a strong advance guard.

And if, against all likelihood, the enemy has had time to be reinforced there, the army is capable of interfering on the afternoon of the 12th and concluding the work begun by Augereau's advance guard.

That advance guard, moreover, is not lost to the front line of battle, and if, during the morning Masséna should encounter much resistance toward Montenotte for instance, it can join and help him. It is a question of 6 or 7 miles.

In any case, the advance guard must protect itself and maintain communications with Masséna to the east, Sérurier to the west, Joubert and Dommartin to the rear. It needs cavalry, and it receives 4 squadrons.

The army is assembled on the 12th, within a small compass, ready in the morning to throw its whole weight against the first objective, Montenotte, if necessary, ready in the evening to turn to the second, Cairo, if that also is necessary. Its action in one case as in the other is directed and prepared by advance guards:

Masséna towards Montenotte.

Augereau towards Carcare and Cairo.

As a matter of fact, Rampon and Laharpe attack Argenteau with 9,000 men at dawn, and break through his flanks. Masséna, who perceives their movement from its start, defeats the two companies, then the battalion which Argenteau has posted at the Bric Castlas and which, arrived during the night, are weakly established. He falls on the rear of the Austrians and precedes them to Montenotte Superieur.

Argenteau, attacked from every side, sees his battalions twirling in bewilderment; he will only collect again 700 men during the day. His reserves, left at Sassello and Squanetto have no time to intervene. They learn of the rout from the troops in flight.

As to Bonaparte, he is at daybreak on the crest by which Masséna is advancing to the north of Altare. From there he sees a few portions of the battlefield; he receives information promptly, and if the action is not proceeding in accordance with his wishes, he can intervene with the troops at hand, if necessary beating back 'Augereau and the other columns. It is soon evident that all is progressing satisfactorily on that side. He will therefore be able to resume with perfect confidence the advance of his army on Cairo.

Unfortunately, he must allow for delays due here to distributions of indispensable things, such as shoes and rifles. The shortage of necessities had indeed been such that in the Augereau division 1,000 men out of· 6,000 had no rifles, to mention only one instance.

Joubert only reaches Altare on the morning of the 12th, Augereau during the day; in the evening Dommartin arrives at Monteffredo.

The disadvantage of these delays is compensated as much as possible by forced marches; the aims of the vari-

THE ECONOMY OF FORCES 85

ous units are modified, but the general plan is being carried out nevertheless.

Masséna is called to fulfill a part of the task which was Augereau's. That very evening he has started the 21st half-brigade to Cairo, the 8th Light Infantry to Biestro.

The distribution of forces, on the evening of the 12th, is therefore as follows:

Laharpe at Montenotte Inférieur, pushing patrols towards Sassello;

Masséna at { Cairo (Headquarters), Biestro;

Joubert at San Donato (between Carcare and Cairo) with outposts towards Casseria and Santa Margherita;

Augereau at Carcare;

Dommartin at Monteffredo, very late;

Bonaparte at Carcare.

The delay as regards Augereau has consequences: Provera had 2,000 men on the heights which divide the two Bormidas, he is allowed time to assemble them. The arrival of Augereau and Joubert in the morning would have enabled the French to carry without difficulty Casseria and Millesimo, instead of which a costly encounter will be necessary the next day, and only the day after that will the French occupy the castle of Casseria.

In spite of all, Bonaparte carries out, on the 12th, the main idea of his plan. He occupies a central position between the Austrians who are retreating on Acqui and the Piedmontese who, ignoring everything, are surprised and still dispersed in their various camps.

On April 13th, Bonaparte will continue the execution of his intentions. " To act towards Ceva, to beat the

Sardinians," principal objective, while he will pursue Argenteau on Spigno, secondary objective. " I am to-day attacking Montezemolo," he writes Sérurier, and he repeats to Laharpe: " It is important to occupy Montezemolo to-day." That is evidently the main purpose for the day. To realizing it he will devote the greater part of his forces; for the secondary action he will use the least possible numbers during the least possible time. To that end, he sends towards Montezemolo:

Sérurier, having Rusca on his right, who will join Augereau about Murialdo;

Augereau following the main road;

Joubert advancing on Castelnovo by San Giovanni;

Dommartin who, if he has passed Monteffredo, will act as reserve to Augereau; otherwise he will rejoin Rusca;

Ménard (with a half-brigade) will remain as reserve at Biestro;

Masséna and Laharpe will proceed early past Dego, send a few companies toward Spigno, and fall back to Augereau's right in order to act toward Montezemolo.

In accordance with these orders, Augereau starts early on the 13th toward Millesimo (having Joubert and Ménard under his orders). Provera was established with 2,000 men on the heights that divide the two Bormidas, connecting the Sardinians of Montezemolo and Austro-Sardinians of Dego. Attacked on the morning of the 13th by the columns of Augereau, he soon loses Millesimo, but locks himself up with 900 men in the old castle of Casseria, at 8 A.M. Augereau seeks in vain, with four small guns and a mortar, to knock down its walls. He seeks in vain to have the defenders come to terms, and at four o'clock in the afternoon he is compelled to attack with the 6,000 men at his disposal. He is repulsed with very heavy losses, and Provera does not capitulate until the morning of the 14th.

But Colli, warned early on the 13th of the attack by the republicans, has gathered forces at Montezemolo. In consequence, while continuing the execution of his plan, Bonaparte is compelled to take Colli into account, and to hold him while overcoming the different obstacles encountered.

On that same day, the 13th, while Augereau was proceeding as an advance guard on Montezemolo, Masséna had been instructed to carry Dego, believed to contain only troops in flight.

Laharpe and Dommartin were coming to Cairo to reorganize a mass capable of maneuvering.

But the 4 battalions which Colli has sent to Dego at Beaulieu's request have reached it on the 12th. Learning there the event of the day, they remain with the remnants of the Rukavina column throughout the 13th. Masséna, arriving in the morning with approximately 2,000 men, learns of the situation from the inhabitants and from deserters; he realizes the insufficiency of his troops for attack.

Such reinforcements as can reach him, Laharpe, Dommartin, are in Cairo only between 11 A.M. and noon. But just then the castle of Casseria is offering unexpected resistance, Colli is displaying strength towards Montezemolo. A reserve is necessary in case he should attack. To that end, Bonaparte keeps Laharpe and Dommartin at Cairo. Masséna will have no reinforcements.

As is already apparent, the simplest actions of the

enemy result in the slowing-up of the boldest leader's advance, and in the conception of a plan which can be carried out nevertheless.

The idea never occurs to Bonaparte of cutting his reserve into two parts in order to simultaneously reinforce the struggle towards Casseria and towards Dego. That would have resulted in weakness everywhere; the attack will continue in one direction, that of Montezemolo, and it will be stopped in the other until further orders.

Meanwhile Masséna has assembled his troops by Rochetta Cairo which he has seized. What will he do there, left to his own resources?

Powerless before Dego, he might have hastened toward the sound of the cannon at Casseria. But it is not his mission, and so he does not go.

He might have been content to take up a defensive position at Rochetta Cairo to prevent the enemy debouching on Cairo. By doing that he would not have prevented the enemy maneuvering in other directions.

He will remain faithful to his mission, which is to act as advance guard on the road to Acqui, and that implies above all the paralyzing of any hostile tentative from that region.

Stopped before Dego and unable to defeat the enemy which holds it, he will attack him there, but by a mere reconnaissance, the result of which will be nevertheless, besides the obtaining of information, to immobilize, to hold these superior forces, and to prepare the attack of them for the morrow. Consequently, about 2 o'clock he orders the reconnaissance of Dego.

THE ECONOMY OF FORCES 89

RECONNAISSANCE OF DEGO (APRIL 13TH)

(*See Map No. 2*)

The 21st half-brigade with 2 guns is ordered up the heights to the north of Rochetta Cairo; numerous patrols are sent forward, take an extended order formation and reach Costa Lupara, Vermenano and the Bric of Santa Luccia; the only 2 guns there are take up positions at Coletto and open fire. The main body of the 21st half-brigade follows in reserve, prepared to support any troops engaged.

In order to push the reconnaissance further to the right, the Rondeau column (approximately 500 men) marches by Massalapo on Gerini.

In order to push the reconnaissance further to the left, Masséna employs the 70th half-brigade which has just arrived with Cervoni. The latter tries to cross the Bromida at the ford by the Bouereu, but is compelled to return to the bridge of Rochetta, where he leaves a detachment while he advances with the remainder of his troops towards Sopravia, preceded by numerous patrols which assume extended formation on contact with the enemy.

All these movements, rifle fire and artillery fire convince the enemy that he is being attacked, and he answers with his own rifle and artillery fire. His fire has small results against the small and extended French troops, but it reveals the dispositions taken by the Austro-Sardinians; it tells Masséna which positions are held.

Renouncing to hold, as in 1794, the heights on the left bank of the Bormida, the Austro-Sardinians have established themselves solely on the right bank. The village of Magliani is about the center of the position, of which the right is at the Bric Rossa, the left at the Bric Della Stella. A first line of defense is formed by the position of La Costa and the castle of Dego.

The artillery of the defenders can be estimated at 16 or 18 guns; they are protected by a few works, barely begun, at the principal points, and a stone redoubt at Bric Cassan (northwest of Magliani).

Such a position is too strong to be carried by sudden attack. In the early hours of the night Masséna, with the approval of Napoleon who has reached the ground, withdraws his troops; he reassembles them at the bivouac of the preceding night, to the south of Rochetta Cairo, and takes up defensive positions there, capable of thus resisting the enemy if he in turn should attack.

Such is a reconnaissance ordered by Bonaparte and by Masséna, in pressing circumstances.

(1) Even for these bold leaders, leadership does not consist merely in charging like a wild boar on the enemy. One must act *with full knowledge of conditions,* proportion the aims and objectives to the means at hand. One must first *reconnoiter.*

(2) In order to reconnoiter one must compel the enemy to *expose* where he is. For that purpose one *attacks* him until his position and frontage are known. But one attacks with the intention of not pushing the struggle; each column will therefore send forward only patrols, scouts who advance, retire, disengage themselves easily when necessary. As means: chiefly long-range action for the same purpose of acting on the enemy without allowing oneself to be grasped, snipers, artillery.

Behind these first-line troops must be main bodies as

THE ECONOMY OF FORCES

supports. The lines of communication and assembly places in the rear are also held (crossing of Bormida, village of Rochetta).

Anyway, the day of the 13th has been lost for the progress of Bonaparte's plan to march on Ceva, and that loss is due to the resistance of Casseria (owing to the lateness of Augereau) and the resistance of Dego. So long as Casseria holds, one cannot attack Dego in force.

The pushing by Augereau of 2,000 men on Millesimo is the sole result obtained on the 13th. Moreover, food is lacking for this army assembled around Cairo.

But on the morning of the 14th Provera capitulates, the much-desired road from Millesimo to Montezemolo is open, one can resume against Colli the execution of the plan prepared, delayed by two days of bad luck but rendered urgent by the army's complete lack of food and supplies, and by its weariness. Yet as long as Dego is not taken there is not the necessary safety from that quarter: between that town and Acqui, Beaulieu can assemble many troops and endanger the whole progress of the army directed on Montezemolo. Dego will be attacked, although the reconnaissance of the 13th has shown the importance of the position, and it may have been further strengthened by additional work or the arrival of more troops.

In spite of the necessity of marching without delay on Montezemolo in the morning of the 14th, the greater part of the army will take the direction of Dego. During that morning, Masséna will be joined by Laharpe, Ménard's brigade, Dommartin and a part of the Augereau division, giving him 18,000 men to be employed before Dego.

Attack of Dego (April 14th)

While keeping the enemy dispersed it is now possible to seriously attack Dego.

Its situation is known by the previous day's reconnaissance; according to the latest reports, no new troops have reached there. Bonaparte then decides to obtain from the numerical superiority which he possesses a complete decisive result: to capture the enemy forces in Dego. For that purpose, Masséna will attack on the right bank in two columns.

The right column of 1,000 to 1,200 men under Lasalcette reaches Gerini by the same road as on the previous day and arrives about 1 o'clock at Bric of Sodan.

Following that crest, the advance guard proceeds to Bric del Caret, reaching it in time to repulse a reinforcing battalion arriving from Squanetto to Dego, two other battalions later. The main body having thus seized with its advance guard one of the roads of Dego, it resumes its advance on Majani, throws back the posts of Della Stella and Del Poggio, and connects with Masséna at the foot of Mount Gerolo, about 3 o'clock.

The left column, under Masséna, has proceeded slowly to allow the others to surround the town; during the greater part of the day it occupies Vermenano, Costa Lupara, keeping up a fire against the enemy by means of numerous snipers and of the two guns at its disposal. The purpose is merely to make a demonstration. At 3 o'clock only, seeing the progress made by Lasalcette, communication established, Masséna rapidly ascends, with his reserve in hand, the slope of Castello, deploys, carries and crosses the locality, deploys again before Costa, and

THE ECONOMY OF FORCES

with Lasalcette assaults Mount Gerolo, the defenders of which fall back on Majani, where they meet the right of the Piedmontese, thrown back also by Laharpe.

Laharpe has maneuvered with the 70th and 99th halfbrigades along the left bank of the Bormida which he has crossed at Rochetta. He is accompanied by 200 cavalry. He proceeds towards Sopravia, leaving 1 battalion to occupy Bormida in order to surround the enemy and protect the artillery which takes up position to the west of Bormida, preparing at long distance the attack of Castello by Masséna. Continuing its progress by Sopravia, the division again crosses the Bormida at the Pra Marenco ford, where it drops 1 battalion, and forms into three columns for the attack of the Casan redoubt, namely:

To the right, under General Causse, 1,500 to 1,600 men by Piano;

To the center, under Cervoni, 900 men by the Bric Rosso;

In echelons in the rear to the left { under Adj.-Gen. Boyer, Chief of Staff of the Division } 800 infantry and 200 cavalry.

The Piedmontese retire from the Casan redoubt on Majani, whence they soon escape by the valley of Cassinelles, seeking to reach the road of Spigno but finding it held by the advance guard of Rondeau who shoot them down from the Bric del Caret. Pursued at the same time by the 200 horse of Laharpe, the enemy's 8 battalions are compelled to lay down their arms; their artillery had not left its positions.

In spite of the urgent need of proceeding to Montezemolo, note what thought and method have been employed in

(1) The use of numerical superiority to obtain a definite result;

(2) The gradual surrounding of the position with a minimum of forces, by the use of such favorable ground as at Bormida, etc.

(3) The decisive blow, prepared patiently, slowly, economically all day, promptly struck between 3 and 4 P.M. by almost all the forces acting simultaneously.

Those will be the constant characteristics of the operations conducted by Bonaparte, Masséna, Lannes.

At the moment when the affair ends, Bonaparte, having already ordered the Ménard and Dommartin brigades to Montezemolo, orders Laharpe to immediately start again for Cairo.

Masséna will hold Dego. But the lack of provisions is complete. His division, the battle ended, disbands to plunder the neighboring villages; it is surprised in the greatest disorder by a hostile detachment, and thrown out of Dego. It cannot be rallied till about 10 o'clock of the morning of the 15th.

As we know, when leaving Voltri on the afternoon of the 11th, Beaulieu has sent 3 battalions under Wukassowitch to Mount Pajole, and thence to Sassello. On the 13th, knowing Dego to be capable of resistance, he has ordered to concentrate there 10 battalions dispersed at Sassello, Acqui, Pacetto, Spigno; 5 of them are beaten on the 14th by Masséna's advance guard to the north of Dego; through a mistake 5 arrive only on the 15th before

THE ECONOMY OF FORCES 95

daybreak. They are the two battalions left by Argenteau to Lezeni and the three of Wukassowitch.

Wukassowitch, who commands the whole, does not hesitate to attack; he seizes Dego and throws back Masséna, whom Laharpe supports. Bonaparte resumes the direction of Dego with all his forces, but only takes the place in the evening, after a fairly lively engagement.

Beaulieu, in spite of the advice and request of Wukassowitch, has sent no reinforcement to the latter. The only battalions he has intact are those of Voltri. He cannot have them before some time.

But that new affair at Dego worries Bonaparte. He fears for the 16th a vigorous counter-attack by the Austrians; reports received make him consider it possible. He orders Masséna to remain on guard for that day at Dego; Laharpe to take up position on Mioglio and Sassello, and to carry out reconnaissances towards Acqui. Headquarters will remain at Carcare.

That means two more days lost, the 15th and 16th, for the execution of the proposed maneuver against Ceva.

But the Piedmontese have abandoned Montezemolo; Augereau occupies it, while Rusca reaches Priero and Sérurier Malpotremo, squeezing thus the camp of Ceva.

On the 16th, Sérurier attacks the camp established on the crest which goes down from the city to Pedagera. He fails. The Joubert brigade on his right gets panicstricken and flees, while the Rusca and Beyrand brigades are thrown back.

On the other hand, reports received by Bonaparte reassure him. The 7 battalions of Voltri have been called back to Acqui where the Austrians are concentrating. The attack of Ceva will be resumed by Sérurier, Augereau and Masséna. The latter, coming for that purpose to Monbarcaro, whence he maintains the separation of the Piedmontese and Austrians, attacks the enemy's left. At the same time, the army is protected by Laharpe who holds Dego and pushes reconnaissances towards Acqui.

On the 17th, the French army finds the camp of Ceva evacuated.

Thus appears, in its execution, the new theory of war, born of the principle of economy of forces and characterized to the highest degree by initiative, attack, action well understood.

(1) Action in *one* direction, that one which is necessitated by the strategic plan, through tactics, that is by the most favorable use of military resources. Therefore, the direction of Voltri being abandoned, one aims at Montenotte, then Dego; Dego being abandoned, Millesimo; Millesimo being dealt with, one returns to Dego, etc.

(2) In each of these directions successively adopted, victory expected from all the forces, or at least nearly all; in other directions, safety assured by troops as weak as possible, intended not to defeat the enemy, but to slow him up, to paralyze him, to reconnoiter him: Cervoni in presence of Beaulieu, Masséna at Dego, Sérurier in presence of Colli.

(3) Constantly, in strategy as in tactics, one seeks a decision by mechanics, by the use against part of the enemy's forces of a main body made as strong as possible by putting in it all the forces which can possibly be spared from elsewhere. That part being destroyed, one aims quickly at another against which one uses again the main

General Foch is shown with General Petain on a visit to the trenches.

THE ECONOMY OF FORCES

body, in order to be always the stronger at the chosen place and chosen time.

As early as 1794, Bonaparte wrote: " The same is true in war as in the attack of a position, fire must be concentrated against one point. As soon as a breach is made, equilibrium is upset and everything else is useless, the place is taken . . . attacks must not be dispersed, but concentrated."

To that end, the forces are constantly maintained as one system:

On the outside, advance guards to (1) attack for the purpose of reconnoitering, of holding the enemy for the main body, or (2) to parry an attack and cover the main body.

Behind, the main body maneuvering in the direction of the objective aimed at.

The main body and advance guards maintain constant inter-communication, so that the whole weight of the mass may at any time be thrown toward the objective sought.

A conversation of Bonaparte with Moreau will, moreover, show us better this new mechanic. The scene occurred in 1799 in the home of Gohier, who repeats it thus:

" The two generals, who had never seen one another, seemed equally pleased to meet. It was noticed that each looked at the other silently for a moment. Bonaparte spoke first, told Moreau how he had long desired to know him: ' You arrive victorious from Egypt,' answered Moreau, ' and I from Italy after a great defeat. . . .' After a few explanations of the causes of the defeat, he added: ' It was impossible for our brave army not to be overcome by so much strength against it. It is always the greater number who defeat the smaller.'

" '—You are right,' said Napoleon, ' it is always the greater number who defeat the smaller.'

" '—But, General, with small armies you have often defeated big ones,' I told Napoleon.

" ' Even then,' he said, ' it was always the smaller number who were defeated by the greater.'

" Which caused him to explain his method to us:

" ' When, with smaller forces, I was in the presence of a great army, I rapidly grouped my own and fell like lightning on one of the wings which I destroyed. I then took advantage of the disorder which such a maneuver always caused in the enemy's ranks to attack him at another point, always with all my forces. I defeated him thus piece-meal, and the resulting victory was always, as you see, a triumph of the greater number over the smaller.' "

IV

INTELLECTUAL DISCIPLINE

Freedom of Action in Obeying

> "I was little satisfied . . . you received the order to proceed to Cairo and did not comply with it. *No event that may occur should prevent a soldier from obeying,* and talent in war consists in surmounting the difficulties liable to make an operation difficult." —Napoleon.

WE have seen into what mechanics, by means of the principle of economy of forces, that ever more rigorous theory of war could be translated: battle with all the forces, used for the greater part as a striking weapon with protective detachments destined to allow free play to that greater part, the whole keeping up constant intercommunications.

We have also seen that the basis of modern war is the use of masses, aiming at a common purpose, or in other words the opposite of independence which would inevitably result in dispersion.

It is evident, therefore, that each one of the units forming the total force is not at liberty to go where it pleases, nor to arrive when it pleases; it cannot be guided by the personal opinions of its chief, however sound such opinions may seem; it cannot act on its own account and seek the enemy or engage him where and when it pleases, even if success should be attained thereby.

Discipline constitutes the main strength of armies.

Armed forces are primarily intended and commanded for the purpose of obedience.

The general commanding in chief can indulge in art, in strategy, all others only carry out tactics, prose. He conducts the orchestra, and they each play their part.

Whether it be a question, therefore, of advance guards or front-line units, of armies, army corps, divisions, brigades or smaller units, every one is a *subordinate* unit.

Every chief of every unit must therefore think of obeying at the same time as he thinks of commanding. Before dictating his orders, he must be inspired by those he has received. To what extent and how? That is what we shall examine.

In war, to obey is a difficult thing. For the obedience must be in the presence of the enemy, and in spite of the enemy, in the midst of danger, of varied and unforeseen circumstances, of a menacing unknown, in spite of fatigue from many causes.

"While dispositions taken in peace times can be weighted at length, and infallibly lead to the result desired, such is not the case with the use of forces in war, with operations. In war, once hostilities are begun, our will soon encounters the independent will of the enemy. Our dispositions strike against the freely-made dispositions of the enemy." (Moltke.)

Then how should we carry out the execution of an order received, unless we preserve our freedom of action in spite of the enemy? The art of war is the art of preserving one's freedom of action.

On the eve of Montenotte, we see

Laharpe ordered to succor the half-brigade of Rampon;

INTELLECTUAL DISCIPLINE

Augereau, leaving Savone at midnight, ordered to fall in behind Laharpe as reserve;

Masséna, leaving Finale in two columns, ordered to reach Montenotte Inférieur;

Sérurier ordered to make a demonstration on Ceva;

Cervoni ordered to hold the road of Voltri;

Rampon ordered to resist at Montenegino.

So many units, so many different tasks; so many separate missions all aimed at a same result: concentration, but always in presence of the enemy and through different means which will rely on the ability of the leaders.

Often the result will be harder to see and to attain. As numbers increase, and simultaneously time and space, the road to follow is longer and harder. At the same time, command, in the narrow sense of the word, loses from its precision. It can still determine the result to be attained, but can no longer specify the ways and means of attaining it. How then can we insure the arrival of these numerous dispersed bodies, except by keeping before them a clear realization of the sole result to be attained, leaving to their initiative the liberty of action to that end? We shall need

Intellectual discipline, primary condition, showing to all subordinates, and imposing on them, the result aimed at by the superior.

A discipline intelligent and active, or rather an *initiative,* second condition for keeping the right to act in the desired direction.

Such must be the embodiment of the military spirit, which appeals to character of course, but also to the spirit, implying thereby an action of the mind, of reflection, and denying the absence of thought, the silent compliance necessary perhaps for the rank and file who need only execute (and yet it is certainly better that they should execute understandingly), but insufficient always for the subordinate leader. He must, with the means at his disposal, *interpret* the thought of his superior, and therefore understand it first, then make of his means *the most suitable use* under circumstances of which he is the *sole judge*.

A leader must not only be a man of character, but also a man capable of understanding and planning for the purpose of obeying.

To the strict, passive obedience of former centuries we shall therefore always oppose active obedience, necessary consequence of the appeal made always to initiative, and of the tactical use of small, independent masses.

And that notion of freedom of action which we find appearing as a protection to our spirit of active discipline, which comes from the necessity of assuring the action of the whole through the combined actions of all participants, we find it also becoming, like the principle of economy of forces, one of the fundamental rules of war.

For in every military operation we have seen that our constant preoccupation is to preserve that freedom: freedom to go to Montenotte, to remain there, to act against Ceva. And, at the end of war, when there are a victor and a vanquished, how will their positions differ except that the one will be free to act and to exact what he wishes from the other, while the latter will be compelled to do and to concede what the victor may decide?

We must be constantly inspired by this idea of freedom to be preserved, if we wish at the end of an operation, and still more therefore at the end of a series of opera-

INTELLECTUAL DISCIPLINE 103

tions, to find ourselves free, that is victors, and not dominated, that is vanquished.

A constant care, at the same time as we prepare or execute an action against the enemy, must be to escape his will, to parry any enterprise by which he might prevent our freedom of action. Every military idea, every plan, must therefore be accompanied by plans for *protection*. We must, as in fencing, attack without exposing ourselves, or parry without ceasing to threaten the opponent.

A historical example will show us what is protection in the larger sense of the word.

On the 4th of August, 1870, the 5th French Army Corps was in the following position:

One of its divisions (the 3rd, under General Guyot de Lespart) was at Bitche with a cavalry regiment;

Two others (the 1st under General Goze and the 2nd under General de l'Abadie) were at and near Sarreguemines, together with 3 regiments of cavalry and 6 batteries of the artillery reserve, also the army corps' transport.

On that evening, General de Failly, in command of the army corps, received from General Headquarters at Metz the following message: " Support with your two divisions the one you have at Bitche." Metz had learnt the defeat of Wissembourg, and had no doubt that the invasion of Alsace by considerable forces would soon follow. It had been decided to reinforce the troops of Marshal MacMahon.

So on the evening of the 4th, General de Failly gets the order: " Concentrate all your forces at Bitche." That is an order which binds his conscience and spirit of discipline. And I must make clear, at this stage, what is meant by discipline, and how it cannot exist without conscience.

To be disciplined does not mean that one commits no offense against discipline; such a definition might suffice for the rank and file, but it is quite insufficient for a leader of whatever rank.

To be disciplined does not mean either that one executes orders received only in such measure as seems proper or possible, but it means that one enters freely into the thought and aims of the chief who has ordered, and that one takes every possible means to satisfy him.

To be disciplined does not mean to keep silent, to do only what one thinks can be done without risk of being compromised, the art of avoiding responsibilities, but it means *acting* in the spirit of the orders received, and to that end assuring by thought and planning the possibility of carrying out such orders, assuring by strength of character the energy to assume the risks necessary in their execution. The laziness of the mind results in lack of discipline as much as does insubordination. Lack of ability and ignorance are not either excuses, for knowledge is within reach of all who seek it.

In execution of the order received, de Failly ordered on the evening of the 4th:

His 1st Division to advance by the main road of Bitche as far as possible.

The 2nd was only to start the next day, and then only in part. It included 2 brigades (Maussion and Lapasset).

To start them on the evening of the 4th was to leave Sarreguemines unprotected. For several days had been

INTELLECTUAL DISCIPLINE

seen, along the frontier, numerous enemy patrols, which had however merely exchanged rifle shots with our scouts. And so one did not believe in abandoning Sarreguemines on the 4th. For the same reasons it will not be abandoned on the 5th. The same reasons will occur again at Rohrbach, at Bitche, and because they prevail the 5th Corps will not arrive. Instead of going to Bitche, de Failly will seek to protect everything; instead of obeying, he will be guided by personal opinions. That is lack of intellectual discipline. Its results will soon be apparent.

On the evening of August 4th, the 1st Division (Goze) had covered 7 kilometers; it bivouacked at the Wissing farm, 2 kilometers from the frontier.

On the 5th, it moved from the Wissing farm to the Fremdenberg farm, 3 kilometers west of Bitche. Having neither advance guard nor flank guard it employed all day in covering that distance of 22 kilometers; it arrived worn out.

On the same day had started from Sarreguemines the brigade of Maussion (from the 2nd Division), the reserve artillery and one cavalry regiment (1st Lancers).

The Lapasset Brigade, of the same division, remained at Sarreguemines to wait until it was relieved by the Montaudon Division of the 3rd Corps. As that division only arrives on the evening of the 5th, the Lapasset Brigade will do no marching on that day. With it are one cavalry regiment and the transport of the army corps.

The brigade of Maussion with the reserve of artillery arrived in Rohrbach at noon on the 5th. It found the country very excited. On the previous evening, a Prussian cavalry regiment had crossed the frontier and approached Rohrbach, after searching the neighboring villages. The 5th Lancers, moving with the Maussion Brigade, had proceeded towards the enemy, who had turned about.

Shortly after the arrival of the brigade at Rohrbach, it was announced that hostile cavalry and infantry could be seen. Part of the brigade immediately picked up arms, and fire had already been opened when it was found that the enemy troops were no other than the 5th Lancers and a detachment of the 68th infantry (from the division of Guyot de Lespart) sent that morning on a reconnaissance.

Under the circumstances, the brigade of Maussion (which was still to proceed to Bitche) did not think it possible to leave Rohrbach. It remained there, notifying General de Failly, who approved the decision taken.

In the reports of the chiefs of the 5th Corps we find the reappearance, with all their hollow importance, of big words, such as the *opening* of Rohrbach, remainders of old-fashioned methods resuscitated by a senseless geography.

An opening, a valley, are not particularly dangerous; there are roads outside the valleys, there are some on the highest plateaus, there are some wherever needed for commerce or the requirements of communication. The road in the valley or on the plateau is dangerous only by the use to which the enemy puts it. If the enemy does not use it, it has no tactical existence, and everything goes on as though it did not exist.

If the enemy were not found, therefore, on the road of the opening within 6 to 8 kilometers (the length of the column) the brigade of Maussion could continue its

INTELLECTUAL DISCIPLINE 107

advance without being stopped at Rohrbach. If he was not within 20 kilometers, it need fear nothing that day.

None of this information is sought; but, because there is anxiety, the brigade will remain at Rohrbach; it will spend the night of the 5th-6th partly under arms.

The brigade, paralyzed by false topographical considerations, halts instead of advancing. It fails to comply with the order. Ignoring protection, unable to safeguard itself, it does not rest, it prepares for the next day exhausted troops.

As to the 5th Corps, the result of these many weaknesses was the following situation on the evening of the 5th:

Guyot de Lespart Division at Bitche;
Goze Division at Fremdenberg;
Maussion Division and reserve artillery at Rohrbach;
Lapasset Brigade and transport at Sarreguemines.

All along the way, instead of the concentration to be carried out reign false theories. Instead of military spirit, of intellectual discipline, we find personal opinions, ignorance of protection. As a result, on the evening of the 5th, the army corps which should and could have been assembled at Bitche spreads out over the 35 kilometers which separate that city from Sarreguemines.

But during the evening the picture changes.

By an order issued at Metz on the morning of the 5th, the 5th Corps was placed under the orders of Marshal MacMahon. The major-general who announced that decision supposed the three divisions of the 5th Corps to have been assembled at Bitche during the evening. Marshal MacMahon telegraphed to General de Failly at 8 o'clock that evening:

" Come to Reichshoffen with your whole army corps as quickly as possible." And he ended by saying: " I expect you will join me some time to-morrow."

Here is another very plain order to be carried out: to arrive as quickly as possible.

General de Failly answers, however, on the 6th at 3 in the morning:

(1) That he can only send, on the 6th, the division of Lespart;

(2) That on the next day, 7th, the Goze Division will come to Philippsbourg (during the 6th it will protect Bitche);

(3) That also on the 7th the brigade of Maussion will come to Bitche;

(4) That the Lapasset Brigade and the transport are at Sarreguemines, definitely cut off.

Entering, as you see, as little as possible into the spirit of the order received, de Failly orders for the 6th:

The Guyot de Lespart Division to start for Reichshoffen;

The Goze Division to remain at Bitche and to be on the 7th at Philippsbourg:

The Maussion Brigade to come to Fremdenberg;

The Lapasset Brigade to remain at Sarreguemines (although the Montaudon Division has joined it there), because communications are cut. It is the enemy cavalry which has cut the railroad at Bliesbrücken.

One has not dared deplete Sarreguemines on the 5th; on the 6th the Goze Division will be kept at Bitche in spite of the series of orders received. We find the same causes prevail as on the day of the 5th; results will be equally disastrous.

On that same day, the 5th, arrives in Bitche, at General de Failly's Headquarters, Lieutenant-Colonel de Kleinenberg, coming from Metz. He announces the presence of a Prussian army corps opposite the corps of General Frossard. And such news, added to everything, again draws General de Failly's attention in that direction.

However, the division of Lespart alone receives the order to start early on the morning of the 6th by the road of Niederbronn, but in consequence of rumors brought by frightened peasants the division delays its departure; it only actually starts at 7.30.

No service of information properly organized. Mere rumors, true or false, generally swollen by fear, are going to dictate military decisions; how could the latter be in accordance with the reality of things?

General de Bernis, with the 12th Chasseurs, precedes the division. There is neither advance guard nor flank guard. Numerous paths debouch on the left of the road followed, and by these General de Lespart fears to be attacked in flank. He only advances one step at a time. The country in front and to the sides is explored by the cavalry, often even by detachments of infantry. Meanwhile, the whole division stops, and it only resumes its advance on the return of patrols with the assurance that one may proceed safely.

Hence occur innumerable halts which the men do not understand. All ranks, excited by the sound of the cannon heard since morning, fret over these delays, and consider the precautions taken as out of place. As they come closer to Niederbronn, wounded men and men in flight are met; they become more and more numerous; they naturally bring bad news; soon they announce the loss of the battle.

When the heights are reached which dominate Niederbronn, the retreating stream of men is seen crossing the town; it is 5 o'clock.

Not until then are communications established between the two portions of the army of Alsace.

Marshal MacMahon orders that infantry division, arrived on the heels of its cavalry regiment to:

Deploy one brigade to the right of the road (Fontanges). And one to the left (Abbatucci). The division's artillery takes up positions.

Before that deployment, the Prussians halt; they have not passed Niederbronn; such is the power of impression caused by the arrival of fresh troops.

The division of Guyot de Lespart had taken from 7.30 A.M. till 5 P.M.—over nine hours—to cover the 22 kilometers from Bitche to Niederbronn.

It brought troops physically and morally exhausted. Above all, it brought *unnecessary* troops. It was too late!

The entire 5th Corps had failed in keeping the appointment.

The battle was lost by its fault.—First consequence.

Could it at least repair the evil done?

Was it to benefit by all the science and precaution employed along the road from Sarreguemines to Niederbronn?

The commander of the 5th Corps, informed at Bitche on the evening of the 5th, of the disaster to the army

INTELLECTUAL DISCIPLINE

of Alsace, was compelled at 7 P.M. to call a conference for the purpose of discussing:

(1) Whether it was possible for the 5th Corps, reduced to 3 brigades (Goze division, Maussion brigade, artillery reserve) to accept battle under the walls of Bitche;

(2) Whether the 1st Corps should be followed in its retreat.

The idea of retreat naturally prevailed. Thus was decided the question of occupying all important points, Sarreguemines, Bitche, Rohrbach; the question of dangerous roads over which it had been impossible to send, because of imaginary dangers, this or that division.

Because they had failed in the battle, in the tactical fact, there remained only danger everywhere.

The troops which it had not been possible to march 30 kilometers to victory had to cover, in a state of demoralization, nearly 100 kilometers (Abbatucci Brigade, from Niederbronn to Saverne) within thirty-six hours.

The 5th Army Corps, without having fired a shot, composed of undoubtedly good and brave troops, was withdrawn from the battle annihilated, depressed, its confidence in the leaders impaired; it was ripe for disaster. In the eyes of the army, and for a long time to come, it was to bear the responsibility for the defeat of Froeschwiller; rightly so, of course, if we confound the command with the troops; unjustly, however, if we judge rightly that battles are won or lost by generals and not by the troops.

False theories, the absence of military spirit, of *discipline* (intellectual and reasoned), complete ignorance of *protection,* of *freedom of action* which protection alone can provide: these were the reasons of the disaster.

On the opposite side, Clausewitz had done away with any inclination to antique methods of fencing; he had preached that battle was the sole argument, and all his disciples had moved to it of their own accord. On our side, we had missed the battle in order to guard strategic points. The commander of the 5th French Corps was not an isolated case in our army; he was only an example of his time and of his surroundings.

Ignorance of protection! Nothing should have prevented the execution of the very simple orders received: " Move to Froeschwiller with all your forces."

It could not have been prevented by distance: there are 55 kilometers from Sarreguemines to Reichshoffen, and three days were available, the 4th, 5th and 6th.

Neither did the enemy prevent it: the 5th Corps has found no trace of the adversary on its way; but everything has been done as though the enemy were everywhere. The Corps should have proceeded *in spite of* the enemy, it did not even proceed in *the absence of* the enemy. False information is accepted without verification even; no scouts are employed; no protection used. It is through ignorance that the 5th Corps has failed.

But the Guyot de Lespart Division is sent. And that division, through lack of reliable information, starts at 7.30 instead of starting at dawn (say 4 A.M. on August 6th); it takes nine-and-a-half hours to cover 22 kilometers which it should have done in five-and-a-half hours. Starting at 4 o'clock, it would have been at Niederbronn about 9; but starting at 7.30, it was there at 1 o'clock in the afternoon. What delays and fatigue in order to arrive so late, after the battle! And if it arrived, that is because the enemy did not show himself. Suppose on the road an enemy of whatever kind, a battalion attacking at some cross-roads. An engagement would have had to be fought on the very route of the advance. Such advance, proceeding already so feebly and painfully, would have ceased completely.

With no enemy encountered, the division arrived too late; if any had been encountered, it would not have arrived at all.

We have fallen far from that use of forces so carefully studied in 1796 by that little army of Italy, concentrated from Savone to Finale with three advance guards at Voltri, at Montenegino, at Ormea, having therefore freedom to act in three directions. Here when action is needed in one direction only it cannot be carried out.

After having seen what was done and the results obtained, let us profit by such costly experience; let us take up the problem again for ourselves; let us place ourselves on August 4th at Sarreguemines, instructed formally to "Concentrate first at Bitche . . . proceed later to Reichshoffen."

What we expect from the principle of freedom of action which will result in protection, is to act in spite of the difficulties of the route, in spite of the unknown, in spite of the enemy—an enemy reported everywhere, existing perhaps somewhere—to *get there,* so that we may carry out the intentions of our chiefs, preserve that discipline which is the main strength of armies. Observed by us, it results in victory on August 6th, it preserves morale in the 5th Corps, it saves an army corps to the French army. That is the art of *knowing* when one *commands*.

How shall we study the question again? With the purpose of being artistic or scientific? Not in the least, simply of obeying, of doing what is expected of us, and with a firm intention of finding in our mind the means of doing it if it be humanly possible.

What is our objective? We are to move to Bitche with the whole 5th Corps, and later to Reichshoffen; that is all. Nothing must sway us from the carrying-out of that order. Let us not speak, therefore, of facing the Prussian troops which threaten General Frossard; of guarding Sarreguemines, Rohrbach, Bitche; those points have no importance for us to-day except as they assist or hinder the advance of the 5th Corps.

The first condition to obeying is, therefore, to visualize above all the order received, and nothing else; then to find the means of complying with it, irrespective of personal opinions and geographical or topographical considerations. These are all quite foreign to the matter.

In order to go to Bitche we must find one or several routes, the safest possible; and as they will never be entirely safe of themselves we must, through special dispositions, provide the safety which is lacking, that is guarantee to the troops the possibility of reaching Bitche in spite of all difficulties.

Taking up the material part of the operation: from Sarreguemines to Bitche we find:

(1) The direct route via Rohrbach, 30 kilometers long, but that route follows the frontier closely; to place

INTELLECTUAL DISCIPLINE

there the main part of the army corps is to render protection impossible;

(2) Another route runs through Zetting, Diding, Kalhausen, Rahling, Montbronn, Lemberg; it is 40 kilometers long;

(3) The route via Sarralbe, Saar-Union, Lorentzen and Lemberg, a total of 50 kilometers.

The first is dangerous, the last is long; it is the second which we must follow with the main part of the army corps, and as it will be a hard one to finish without a stop we must start on it the evening of the 4th. In order not to lengthen the column, we shall send by the 3rd route the transport, which is of *no combatant value.*

The army corps, starting on the evening of the 4th by the Zetting and Diding route, will bivouac at the end of the first day's march in Wittring and Achen. The next day it will reach Bitche without difficulty.

But we must also be sure that the enemy will not interfere with the move. That result can only be attained if, during these days of the 4th and 5th, he does not appear (*which depends on himself*), or if, appearing he is kept at a distance (*which depends on us*).

The first thing to do is, therefore, to know if he appears, if he shows himself in the territory which we shall cross, or on the flank of the route which we intend to follow. That necessitates *information,* a service of information by the cavalry. And where shall we seek the information? In every direction by which the enemy may arrive to reach our route, on every dangerous road.

Such information, sent from a sufficient distance, from 20 kilometers for instance, affords ample protection if it be negative. But can information be obtained at such distances? Not in this case, for a safe margin of 20 kilometers would have our cavalry operating towards Bliescastel and Deux-Ponts, right in enemy country.

Moreover, the information may be affirmative. Instead of verifying the absence of the enemy, it may establish his presence within a radius of less than 20 kilometers, and it may be too late to parry it unless, through previous dispositions, the case has been foreseen.

From all this we see that at the same time as we organize the service of information by cavalry we must foresee the case of the enemy being found less than a day's march from the column. To continue to insure its freedom of advance, we must then interpose, between the route followed and the enemy, a resistance capable of holding that enemy during the time necessary for the passing of the column. The army corps camping at Achen, Kalhausen, Weidesheim and Wittring will be protected, on the 4th, by a flank guard composed of 1 brigade, 3 batteries, 1 regiment of cavalry, and established at Woelfling and Wiswiller, junction of dangerous roads.

The army corps will also order the occupation of Singling by a detachment of the regiment of Achen.

The Woelfling brigade will, in turn, protect itself by a system of pickets to include:

At the farm of Hemscapel:	1 company	the Supports occupying
At the northern end of the wood:	1 company	positions on all dangerous
At Gross-Rederching:	2 companies	roads.

Further back, a reserve (equivalent to a battalion) ready to move to any threatened part of the line, and

INTELLECTUAL DISCIPLINE

placed for that purpose at the crossing of the dangerous roads.

In advance of that scheme of *safety through resistance*, a scheme of *safety through information*, composed of:

1 troop of cavalry at Bliesbrücken;
1 troop at Rimling;
1 troop at Obergailbach;
1 squadron at Rohrbach.

The remainder of the cavalry regiment being encamped with the main body.

On the next day, the column, resuming its march early, has 25 kilometers to cover in order to reach Bitche, its head starting from Kalhausen. That can be done without fatigue.

If the column starts at 5 A.M. (it is the 5th of August), its head is at Bitche six hours later, that is at 11.15; the tail-end (16 kilometers behind) reaches there four hours later, at 3.15; allowing for a long rest, the whole column is assembled in Bitche at 4.15.

But that necessitates not being interfered with by the enemy. To that end, the flank guard must successively occupy dangerous roads as the column progresses.

The flank guard must be in Rohrbach, therefore, when the column passes Rahling, at Petit-Rederching when the column passes Enchenberg.

How long shall the flank guard remain in Rohrbach, considering that the column is not one point, but has a length of 16 kilometers?

As there are 8 kilometers from Rohrbach to Rahling, the flank guard can leave Rohrbach when there only remain to pass through Rahling 8 kilometers of the length of the column. It is evident that any enemy who, after that time, passes Rohrbach will strike nothing at Rahling. As to the column, its head leaves Kalhausen at 5 o'clock, arriving in Rahling at 6.30 and at Enchenberg at 9.30, while the tail passes these points four hours later.

The flank guard must then be in Rohrbach at 6.30, leaving Woelfling for that purpose at 4.30. But, as we have seen, two hours later, namely at 8.30, half the column has passed Rahling; there only remains the other half, 8 kilometers long, and Rohrbach can therefore be abandoned without danger. As a matter of fact, the flank guard can abandon it an hour sooner, thanks to its service of information by the cavalry.

If, for instance, at 7.30 the cavalry sends from Bettwiller the information that "there is no enemy at Bettwiller," we are sure that there can be none in Rohrbach at 8.30, since there are 4 kilometers between these two points; we can therefore abandon Rohrbach at 7.30.

The flank guard will then move to Petit-Rederching and Halbach, holding the dangerous roads which lead to Enchenberg, until there only remains to pass at that point a length of column equal to the distance which separates Enchenberg from the point occupied by the cavalry. In order that the flank guard may maneuver in such manner, it must reach in good time the defensive positions, that is the crossings of dangerous roads, and to that end it must be closer to them than the enemy.

It will reach Rohrbach if, on leaving Woelfling, it has information that the enemy is not in Rimling, because the distance from Rohrbach to Rimling equals that from Woelfling to Rohrbach.

In the same way, after leaving Rohrbach it will reach

INTELLECTUAL DISCIPLINE

the next cross-roads if it has information that the enemy is not in Bettwiller, and the next cross-roads again if the enemy has not reached the height of Hottwiller.

As can be seen, the work of the flank guard will be guaranteed by a service of information reconnoitering along its route, and operating at distances equal to those of the roads to be reached. To-day, therefore, the main body of the flank guard's cavalry will pass through Rimling, Bettwiller, Hottwiller, etc., and an advance guard of one battalion with a troop of cavalry will precede the column.

With such dispositions we are assured of marching without a pause our main body.

If the enemy appears, he will always find on the route some force to interpose and capable of serious resistance, capable at least of lasting, and thus absorbing the enemy's activity at a considerable distance from the main body, whose advance is thus protected.

In reality, on August 5th, 1870, there was no enemy in the region; the 5th French Corps has seen no sign of any. Its head would, therefore, have reached Bitche at 11.30, its tail at 4.30.

The marshal's order would have found the head of the column (one regiment) in a condition to march further—it had only covered 25 kilometers. It would have been pushed on to Engelsberg, 6 kilometers from Bitche, but still in advance of that point would have been pushed the Lespart Division which had done no marching all day. It would have proceeded to Philippsbourg (14 kilometers), throwing out a flank guard of 1 regiment, 1 battery and 1 squadron towards Sturzelbronn and Main-du-Prince.

The advance would have continued the next day, August 6th, without difficulty; there is only one dangerous route in the morning, that of Sturzelbronn; it was held by a force capable of holding out a long time, disposing of 8 kilometers over which to retire into a defile of mountains and woods.

The march, resumed at 6 o'clock from Philippsbourg and at 5 o'clock from Bitche, would bring the head to Reichshoffen (10 kilometers) at 8.30, the tail at 1.30.

The arrival of the 5th Army Corps at Reichshoffen in good time was therefore possible; it could have been accomplished even if the enemy had interfered on the flank of the route followed, but on condition of using:

Activity of mind, to understand the purposes of the Higher Command and to observe the spirit of these purposes;

Activity of mind, to discover the material means of fulfilling them;

Activity of mind, to fulfill them in spite of the enemy's efforts to preserve his freedom of action.

On condition, in brief, of displaying proper *discipline*.

V

PROTECTION

DRAGOMIROW has written: " The principles of the art of war are within reach of the most ordinary intelligence, but that does not mean that it is capable of applying them." Instruction would therefore be useless which merely pointed out these principles without going into their application.

For that reason we propose to now study in detail the work of the 5th Corps' flank guard.

We have seen how forces were divided into a *main body* and *protective troops,* in order to allow the army corps to reach Bitche in spite of the enemy; how it was intended to thus carry out the order given, by opposing any intentions of the enemy.

We have also seen by what estimates of time, what use of information, the main flank guard succeeded in being, at the proper time, at dangerous cross-roads.

Let us see to-day how, the opponent arriving, the protective troops will operate; what tactics and dispositions they must adopt in order to carry out their purpose.

Let us examine whether these conditions adopted for protection, which we have considered *necessary* are effectively sufficient.

I have supposed, for that purpose, an enemy force, estimated at not less than a division, encamped on August 4th towards Alt-Altheim (6 kilometers from the frontier, 12 from Bettwiller), and starting the next day with the intention of holding to the mountains the 5th French Corps, of preventing its junction with the forces of Alsace. The adversary also, by pursuing the concentration of his forces towards the Sauer, seeks to maintain the dispersion of ours.

I have shown that, under these circumstances, with an enemy attack, the 5th French Army Corps of 1870 could not have arrived, that it could have brought none of its forces to Bitche.

Let us see how the same army corps, protected as we have planned, could have arrived.

In order that it may have evacuated Enchenberg, it needs (at 4 kilometers per hour):
hours' marching.
$$\frac{6+13+16}{4} = \frac{35^k}{4} = 9 \text{ hours' marching.}$$

From Alt-Altheim to Enchenberg, there are $\frac{22^k}{4} = 5$ hours' marching. Without protective troops, therefore, there would surely be an engagement at Enchenberg.

The enemy division, the head of whose column has left Alt-Altheim at 5 o'clock, reaches Bettwiller (12 kilometers distant) at 8 o'clock, if nothing has stopped it. The flank guard, on the other hand, has reached Rohrbach at 6.30. Already then, its cavalry regiment is in Rimling, with pickets at Erching, Guiderkirch, Epping. The picket of Guiderkirch, soon pushed back by enemy squadrons, is reinforced by the cavalry regiment, which opposed the enemy's advance on Guiderkirch and Moulin de Rimling; it warns the commander of the flank guard, and covers the direction of Bettwiller.

PROTECTION

The presence of this cavalry regiment compels the enemy, who has no greater cavalry, to call on his infantry advance guard to clear the way. It is necessary to start maneuvering. That advance guard, to reach Bettwiller, needed $\frac{12k}{4} = 3$ hours' march (8 A.M.), without enemy encounter. The hostile cavalry interfering, it will only reach that point after 8 o'clock. What has the flank guard done meanwhile?

The flank guard started from Woelfling at 4.30, its head has reached Rohrbach at 6.30.

On arrival, it has protected the occupation of the ground by pushing 2 battalions to Hill 376; it has retained at the same time the possibility of acting further, of continuing its advance to the east, by means of an advance guard (1 battalion) pushed to the station of Rohrbach. These precautionary measures would allow it to make a long stay in Rohrbach without inconvenience.

But, as we have seen, Rohrbach can be evacuated at 7.30, if by that time the enemy has not reached Bettwiller, which we shall know by our cavalry. On the other hand, the brigade takes an hour to pass by: the head arriving at 6.30, the tail will pass at 7.30; Rohrbach would be held, therefore, until 7.30 even without halting the column. But let us suppose that for the sake of greater safety a half-hour's halt is made there.

The column will assemble to the north of the locality, its head arriving at 6.30 and leaving at 7, while its tail arrives at 7.30 and leaves at 8.

From the officer in command of the cavalry regiment information is received between 6.45 and 7 o'clock that the enemy (cavalry first, infantry later) has been seen on the Peppenkumm-Guiderkirch road, and that he himself, if compelled to retire with his squadrons, will retire on Bettwiller. That information confirms the necessity of blocking without delay the Bettwiller-Petit Rederching-Enchenberg road.

On the other hand, as time goes by, the occupation of Rohrbach becomes less imperative; its evacuation can be prepared.

Because of this double reason, as early as 7 o'clock, the brigade is moved from Rohrbach to 800 meters southwest of the station, on the main road (at the crossing of the path which leads up Hill 376); the assembled troops march in mass formation, the troops in column of route formation retain it. The artillery is on the road, covered by the 2nd Battalion kept at Hill 376, by the 1st pushed to Petit-Rederching (occupied at 7.15), that is by two advance guards maintaining until further orders the possibility of acting in one or the other direction.

At 7.45 the whole brigade has reached the objective. The 2nd Battalion, on Hill 376, receives the order to rejoin.

But meanwhile, the commander of the 1st Battalion, arrived as advance guard in Petit-Rederching at 7.45, has prepared his plan. What is his mission, as advance guard commander for the brigade? To prepare the entry into action of that brigade against an enemy appearing from Bettwiller. What is necessary to the brigade for that purpose?

The space necessary to the employment of its forces;
The time necessary to their arrival and deployment.

PROTECTION 125

In order to fulfill that double purpose, he occupies with troops all the space necessary, at points where they may hold the time necessary. In this instance: Petit-Rederching, Hill 349, Hill 353.

One company is moved to each of these heights. The two others take up positions in Petit-Rederching, which is placed in a state of siege. At 7.30 the brigade continues its move, still protected on the Petit-Rederching-Bettwiller road as shown. It is also protected on the Bettwiller-Rohrbach road by the battalion from Hill 376, which becomes a rear guard and retires along the crest, as soon as the main body has finished leaving the second assembly point.

We thus find our main body still keeping the possibility of action to the right or to the left, by maneuvering under protection.

At the same time, the cavalry has been ordered to continue delaying the progress of the enemy, to reconnoiter the strength of his column, to watch over the roads east of Hottwiller, etc.

The brigade assembles to the south of the Petit-Rederching cross-roads.

Information accompanies it there. About 7.30 an officer's patrol has seen a column of infantry and artillery entering Guiderkirch on the way to Bettwiller. Also, after 8 o'clock infantry has been seen to occupy Bettwiller. The hour shows us that now (after 8 o'clock) the road of Rohrbach is of no interest, while on the other hand we must at once close that of Petit-Rederching, and later on those further east.

How shall we close the first road, how shall we halt the enemy? Carnot has already answered that question:

"The enemy will not fail to detach some troops on your flank to stop you. You must place opposite these troops a division which, whether by its strength or by a commanding position, will disperse them or stop them."

Therefore, use force and, lacking force, use position, which is sufficient for we are not asked for victory, but merely to hold the enemy while the column goes by.

Let us estimate that time: The head of the army corps will reach Enchenberg at 9 o'clock; four hours for passing; the tail will leave it at 1 o'clock in the afternoon.

From Enchenberg to Petit-Rederching there are 6 kilometers, or an hour and a half of marching.

To let the enemy be there at 1 o'clock he should leave Petit-Rederching at 11.30, so that he will not get there if he be kept at Rederching until 11.30, or at points nearer Enchenberg until later. The commander of the flank guard must find means, therefore, of holding out until 11.30 at Petit-Rederching, or until 12.30 at Heiligenbronn. It is now after 8 o'clock.

How to hold out? If the enemy be not in strength, the question is not difficult to answer. If the enemy is more powerful, use strong positions.

Impregnable positions do not exist, for any position used merely for passive defense falls by maneuver.

It is like a fencer who merely parries; sooner or later he is hit. It is like a warrior placing all his trust in his armor: there is the defect in the cuirasse which is always found eventually.

But, lacking impregnable positions which do not exist, there are, especially with modern weapons, many strong positions.

Because of their power, modern weapons forbid any

maneuver under fire; because of their range, they compel assuming at long distance battle formations, deploying far away; because of their rapidity of fire these necessities may be enforced even by troops comparatively weak.

A position occupied delays, therefore, the enemy, on condition that it be a real position. What is needed to that end? What is understood by a position, in the modern sense of the word? It is ground suitable to defense based on fire and strength; it is a place which offers:

Positions from which one sees far and can shoot far— clear fields of fire;

Obstacles to enemy advance.

When these conditions are met, the enemy is compelled to maneuver from afar, until the last moment (arrival on the obstacles), to engage all his resources, artillery and infantry, which means advancing painfully, wasting time while he would wish to proceed. The points to be occupied are here Hills 349 and 353, and Petit-Rederching.

But such a position will finally fall after a certain time which may not suffice for the passing of the army corps. Whatever additional time is necessary must be obtained from a second position. While organizing the first position, we shall have sought and prepared a second one.

The result is to organize the action in depth, to prepare a series of successive encounters, in each of which decision will be avoided, so that only a part of the troops will be engaged. We get the use of forces governed by the special circumstances in which we find ourselves.

On the first position what shall we put? No premature deployment; it might be unjustified, the enemy having given no indication yet of the point which he intends to strike. The greater part of our forces would fail us if we used deployment too early.

What we must do, is to *occupy the position* with troops which allow of bringing there all or part of the main body, as may be deemed necessary, when the time comes, when the attack has shown the point it seeks to overcome.

Occupation by what kind of troops? By such troops as can use fire at long range, for our intention is to compel the enemy to as long a maneuver, over as long a distance, as possible. In the first line we shall therefore place some infantry and all the artillery. Following that idea, we shall support the advance guard with

1 Regiment having
- 1 Battalion at Petit-Rederching, having:
  - Northern edge of Petit-Rederching: 1 company;
  - Hill 349: 1 company.
- 1 Battalion at the northwest point of the wood, with 2 companies on Plateau 353.
- 1 Battalion in reserve to the south of Petit-Rederching.

That position will not be attacked to the same extent from all sides. The blow will probably be aimed at its extremities, or at least at one of its extremities. A reserve must prevent this: the reserve of the 1st Regiment will keep itself in readiness to support the flank attacked. If it be the right flank, the 1st Regiment will occupy the northeastern edge of the wood. If it be the left flank, the Tuilerie and station can be occupied.

The artillery is ready to take up positions on 353, appearing only to open fire, endeavoring to control as far as possible the Bettwiller road, in order to prevent its use by the enemy. The remainder of the brigade is assembled to the south of 356.

The cavalry is exploring the east, where some dangerous roads remain, and for that reason it occupies the ravine of Nonante.

The front of the position measures 1,500 to 1,800 meters, which, for one regiment reinforced by artillery, is very acceptable under present defensive conditions.

Should this first position be lost, a second will be needed, commanding the Petit-Rederching-Enchenberg road.

The station, 356, the northern edge of the wood and Halbach offer a second position, the reserve being to the west of Siersthal.

A front of 3,000 meters would evidently be a great deal, if we intended to resist firmly with our one brigade, but the position includes 1,500 meters of a wood which can easily be held with small forces.

Moreover, there is a great need of holding Halbach, through which the enemy might take the direct road to Bitche, or the road to Enchenberg via Siersthal.

Again, a frontal attack against the northern edge of the wood can give no results; the enemy will seek to extend to the right or left of the wood, and we can then oppose to that attack the greater part of the forces (the reserve) if we have at first held the position but lightly.

At the proper time, this occupation of the position might be as follows:

2nd Regiment { Halbach, The wood, Hill 356;

1st Regiment (leaving 2 companies at the station) reforming as General Reserve west of Siersthal, at the crossing of roads leading to { Hill 356, Halbach, Siersthal; } { 3 directions in which it may have to act;

Artillery south of Hill 356;

Cavalry searching always east on the Hottwiller road; but as cavalry would be handicapped in fighting there because of the wooded nature of the ground, the greater part of the regiment can be moved to the left, where is found more open ground.

As we see, a front line on: Station, 356, Wood. No more than on the previous line (349, Petit-Rederching, 353) do we engage on principle all the troops, because we are not at this time seeking any *decisive* engagement. None is necessary, and it might be beyond our power.

We are merely seeking to delay the enemy by compelling him to maneuver.

To what extent shall the forces affected to each line of defense be used up there? It depends on circumstances.

It is evident that if, against a lightly held line, the enemy stops, reassembles, slows up, maneuvers cautiously, we do not engage the whole of the 1st Regiment on the first line; its 3rd Battalion can be sufficient for holding, in that case, the Station, 356, the Wood, when the time comes. If, on the other hand, the enemy at once deploys important forces and begins a vigorous attack, the whole 1st Regiment may be used up in retarding it; the 2nd Regiment will undertake to occupy the second position.

If we wish to figure in hours, certain results appear inevitable.

The enemy column, bringing its advance party to Bettwiller at 8.30, comes under fire of the guns at 353; it can only keep advancing under natural cover of the ground.

The hostile advance guard, beginning to maneuver, seizes and occupies 353 (Kleinmuhle), Hielling and Bettwiller, its reserve being held at the last-named place.

From there it undertakes a reconnaissance which will show it an extended front and artillery; from there also it will protect the column while it arrives and assembles.

As a division in column takes about two hours to pass a given point, it is not till 10.30 that the last man arrives. Until now, the assembly has been east of Bettwiller.

It is to be presumed, therefore, that the line Petit-Rederching-353, even if feebly held, cannot be carried before 11.30. The Station will be carried about 12.30.

If the enemy, continuing to maneuver along his right, advances along the Petit-Rederching-Enchenberg road, the reserve regiment deploys and occupies Heiligenbronner Wald, the Heiligenbronn Farm and the southern point of the wood. The artillery is at 372, behind which position is assembled the 2nd Regiment (south of 372, for instance).

But there are 6 kilometers from Bettwiller to this

position; the enemy, once he has extended, will have spent over two hours to cover them. The attack can only occur after 1 P.M. before Heiligenbronn; at that time the army corps has passed Enchenberg, so that the flank guard can give way and retire on Siersthal.

If the enemy, having taken Petit-Rederching and 353, instead of continuing his movement along the Petit-Rederching-Enchenberg road, undertakes the attack of Halbach, he comes to an angle of the wood where he cannot take advantage of his superiority in numbers; his march is inevitably delayed by the broken-up nature of the ground. The 1st Regiment reinforces that point, sends to it about one battalion, and moves the remainder to the east of 377, passing through Siersthal and occupying it. The artillery also comes to the east of 377.

There the length of an engagement may be great.

At nightfall, the brigade can retire towards the Briqueterie, 416 and the Fremdenberg Farm.

The example given shows clearly how flank guards of large bodies operate: their tactics consist in constantly maneuvering to move from one road to another, to fight a rear guard engagement, etc.; they are mobile troops, who seek and use successively the occupation of field positions.

The wooded region we encountered favored direct defense of the roads along which the enemy appeared. Under less favorable circumstances, on open ground, he might make a feint on one road, seizing another ahead of us without thereby losing his cohesion.

He must be stopped nevertheless; defense being no longer possible, there is no other resource than to attack.

The case would also have occurred if the enemy, by a

night march, reached Rohrbach for instance or Petit-Rederching ahead of the flank guard. How could he be held? Once again, by attack.

This shows that a protective duty does not necessarily entail a defensive attitude; it is often by offensive that it can best be fulfilled. The duty to fulfill is a thing quite distinct from tactics. It is from the consideration merely of the objective, under existing circumstances, that we can decide on the tactics to be observed.

A historical example will show this clearly: the example of Kettler's Brigade before the army of Garibaldi. (*See Map No. 3.*)

In the evening of January 20th, 1871, the German Southern Army's Second Corps has its advance guard at Dôle, main body at Gray, transport at Thil-Chatel. The Seventh Corps, further north, has crossed the Saône.

A flank guard (half of the 4th Division) is in Essertenne.

That same day, January 20th, arrived in Turcey and in Saint-Seine Kettler's Brigade of the 2nd Army Corps. It had been previously kept back to cover the Chaumont-Châtillon-Montbard railroad which supplied the Second Army, and to protect the baggage of the 2nd Corps. At Saint-Seine-l'Abbaye it received the order to seize Dijon, or rather to immobilize the French forces in Dijon.

What was the task at Dijon of this brigade of 4,000 rifles and 2 batteries? At Dijon had been assembled the Army of the Vosges, under the orders of Garibaldi, and the Pélissier Division. The whole amounted to from 30,000 to 50,000 men.

These figures seem very doubtful at German General Headquarters, considering the complete inactivity of these troops, which not only had not interfered in the Southern Army's progress across the Plateau of Langres, but which had even abandoned the Saône bridges to the heads of the German columns, without any serious engagement. However, on January 18th and 19th, German General Headquarters was informed of the arrival at Dijon of siege guns and of important reinforcements, additional means for Garibaldi of interfering in the extension of the Southern Army, besides hindering any offensive maneuvers which that same army might undertake in the basin of the Saône.

General Manteuffel, while pursuing his plan of acting with all his forces against the principal enemy army, that of General Bourbaki, was therefore compelled to protect himself against such important forces as were at Dijon, to forestall their attacks.

He saw to it on the 20th by means of half of the 4th Division kept as flank guard at Essertenne. To recover later the free use of this force, he intrusted the duty, beginning on the 21st, to the Kettler Brigade, specially called to Saint-Seine and to Turcey.

From that brigade the following are missing: 2 companies left in Montbard to guard the railroad, and 1 battalion and 1 squadron which, having escorted all the convoys of the 2nd Corps, find themselves on January 20th at Is-sur-Tille.

The brigade has therefore at its disposal on that day:

At Is-sur-Tille { 1 squadron
{ 1 battalion (of the 61st)

At Saint-Seine-l'Abbaye
{ ¼ squadron
1½ battalion (of the 21st)
1 battalion (of the 61st)
2 batteries }

At Turcey
{ ¾ squadron
1 battalion (of the 21st)
1 battalion (of the 61st) }

What is happening in Dijon?

Garibaldi has allowed the Southern Army to cross the Plateau of Langres without threatening it in any way.

On the 19th, while the enemy crosses the Saône, he begins to move the Army of the Vosges, intrusting to General Pélissier the protection of Dijon. He leads his troops, in 3 columns, to 7 kilometers north of Dijon, and halts on a height near Messigny, from which he watches the flank guard of the 2nd Corps (half the 4th Division) march by. By pushing only to Thil-Châtel he could have met the tail of the column of the 2nd Army Corps.

Even on the 20th, he would still have prevented the passage of that army corps' convoy. Instead of acting, he is content to return to his encampments around Dijon.

By acting as he did, he left open not only all the roads of the Côte d'Or, but also the crossings of the Saône.

As to Dijon, it has been fortified, but the Germans are only dimly aware of it.

Talant and Fontaine-lès-Dijon have been strongly organized, and armed with large caliber guns, some of which command the road of Saint-Seine and the road of Langres. In the same manner, La Filotte, Saint-Martin and La Boudronnée have been strengthened for defense and connected. The village of Saint-Apollinaire, on the road to Gray, is joined to the works begun on the south side by the Germans and continued by the French. To defend these positions, the Government of National Defense has organized the Pélissier Division; it intends to use the Army of the Vosges for campaigning purposes. But that army is still at Dijon, and it is against the possibility of an attack on its part that General Manteuffel protects himself by ordering General Kettler to march against Dijon.

How will the latter fulfill his mission: to immobilize in Dijon the Army of the Vosges, or at least to keep it from interfering in the events which are going to occur on the Saône, and later on the Ognon?

He must, *without delay,* encounter the enemy, and for that purpose attack him. There is no time to assemble. From Is-sur-Tille to Turcey there are nearly 40 kilometers; there would still be 20 if the meeting were halfway; that means a day lost. There only remains to march on the enemy as rapidly as possible. No glorious decision is aimed at, that will come on the Ognon. At present the only purpose is to immobilize the opponent.

So we see, on the morning of the 21st, three columns which altogether do not exceed

5¼ battalions
2 squadrons
and 2 batteries

proceed towards Dijon as follows:

Center: from Saint-Seine to Dijon, { ¼ squadron
column of Kettler   2½ battalions
                                       2 batteries

Right: from Turcey to Dijon, Major Kroseck { ¾ squadron, 2 battalions
Left: from Is-sur-Tille to Dijon, Major Conta { 1 squadron, 1 battalion

The center column meets French irregular forces on leaving Saint-Seine, and again at Val-Suzon; it reaches the Changey Farm at 1.30, and is greeted there by artillery fire from Talant and Fontaine.

The 1st Battalion occupies the heights to the right and left of the road; the two batteries take up positions under cover at 390. Daix is attacked and easily taken.

Such is the opening action of an advance guard whose task is to prepare that of the main body by:

(1) Reconnoitering the enemy;

(2) Covering the preparations of the main body (arrival, assembly, deployment);

(3) Engaging the opponent.

Here the reconnaissance is made. The opponent displays his masses on the slopes of Talant and of Fontaine. Only the two other operations remain. To cover over a wide front and with few troops, strong points will be used, that is points where one can last, where the insufficiency of numbers is counterbalanced by the difficulty of the obstacles or by the effect of the fire. In this case, Daix and the two spurs which command the road. The occupation of the dominating points and of the strong points by the advance guard's infantry will be consolidated by that advance guard's artillery. The battery which moves with the advance guard takes up positions.

Taking up the third operation, the engaging of the opponent, he will be immobilized by the threat of an attack; there is little infantry, and the ground is well covered by hostile artillery; there will be little use made of the infantry, artillery especially will be used, as it can last as long as ammunition holds out; besides, the Prussian guns are superior to the French guns, let them be used; the battery of the main body comes to reinforce the advance guard's artillery. The greater part of the infantry does not appear in the firing line. Meanwhile the French observing that cautious attitude, launch an attack on Daix. They fail; they seek to maneuver, to attack it from a flank. The obstacles of the strong point render all their efforts useless.

The Kroseck column (right) arrives only about this time, although it has met no difficulties on the way. It carries Plombières. Communications with it are established. General Kettler then decides to pass to the offensive. What shall he attack? One point, Talant, because it is against this point that he can act with a maximum of forces, because the slopes of the valley provide to the troops of Major Kroseck in particular some good cover. The attack is prepared by the fire of the two batteries. The Kroseck column (2 battalions) and 1 battalion from the center column are thrown against Talant.

The attack wins all the ground to the foot of Talant; it fails before the village. It is 6 o'clock and dark; it is necessary to stop. But in order to complete a success *which has not been achieved,* in order to immobilize the opponent, he is kept under the threat of an attack imminent at any time; General Kettler orders the troops which have attacked to spend the night at the foot of Talant, a few hundred meters from the position.

It is January 21st. The weather is very cold.

It has not been possible to establish communications with the detachment of Is-sur-Tille; it becomes necessary to guard in that direction against enemy action: 1 battalion is detached to Hauteville. The remainder of the column remains opposite the enemy.

As to the column of Is-sur-Tille, it has been unable to either carry Vantoux or to establish communications with General Kettler. It is threatened from all sides, and has retired on Savigny-le-Sec.

In spite of the audacity displayed in the enterprise, notwithstanding all the energy shown in its execution, the day has been only partly successful.

The attack of Talant has been costly: the brigade has lost 19 officers and 322 men.

The night in the open exhausts the brigade. There has been no wood or forage. For food, there was only a little bread, biscuits, lard. Towards the end of the night, snow begins. General Kettler bivouacs near the Changey farm; the buildings are filled with the wounded; the General spends the night on the road, protected only by his two horses against the wind that blows on the plateau.

The enemy has shown himself in considerable strength, he has resisted everywhere. And if he has resisted everywhere in force it means he has large numbers. He will claim a victory undoubtedly, but one more victory of this kind will mark the beginning of the final disaster. The purpose sought by the Kettler Brigade will be accomplished.

Besides, the influence of the battle has been felt at

Dijon. Fearing another attack, the municipal authorities have asked that their city be spared the sufferings of a bombardment. However, on the next day the Kettler Brigade was in a state which forced it to go into rest camps in order to recuperate and to feed its men, which was becoming very difficult in a naturally-poor country, pillaged by the Garibaldians.

Efforts must also be made to save from destruction, and in any case from isolation, the Conta detachment. Finally, the ammunition is running short. An ammunition section is called.

Therefore, on the 22nd, rest. But the food sought is lacking everywhere. Of course, observation is maintained. The French attempt an attack on Plombières, and are going to carry the place when the Prussian officer in command has the civilian inhabitants dragged from their cellars, and stood in a living barrier before which the French fire ceases. Prussian conception of the principle of protection.

On the 23rd, in order to find villages naturally richer or less drained of supplies, General Kettler decides to move his brigade, by a flanking movement, from the hill to the plain.

The movement is carried out from Hauteville by Ahuy on the Valmy farm; the brigade is assembled at 11 o'clock, having joined the Conta column. The news arrives as follows:

(1) Bellefond and Ruffey, occupied the previous day by the enemy, are abandoned;

(2) The flanking movement of the morning has suffered no interference whatever from the enemy, not-

withstanding the short distance at which it was carried out;

(3) Peasants and prisoners add that many troops have left for Auxonne.

General Kettler does not hesitate. His brigade is in very poor condition; the position of Dijon is particularly strong. A first check has been suffered, he will suffer another, yet he does not hesitate. The enemy is maneuvering, disappearing, he must be *held back,* and for that purpose he must be *attacked.* In order to attack, one must first see clearly, one must know where the enemy is, what positions he occupies, one must recognize what point to attack with the main body, which must not be thrust blindly anywhere, in any way.

Such reconnaissance is the first duty of the advance guard, duty to which is always assigned a minimum of forces. A battalion of the 21st is ordered, at 1.30, to sweep back the irregular troops from the heights north of Pouilly. They are swept back, and behind that line the assembled brigade advances. Simultaneously, the cavalry patrols which surround and protect the flanks of the advance guard report that considerable enemy forces are leaving Varois and Saint-Apollinaire, marching towards Ruffey.

The service of information, which has sufficed until now to guard the flanks of the attack, is no longer able to stop the oncoming enemy. Resistance must be made strong enough to be able to last: 1½ battalion and 1 squadron are sent to Epirey.

At the same time, moreover, as Epirey is occupied, one company of the main body is dispatched to Ruffey. While action is prepared, protection is assured; such is the constant preoccupation: once the attack is launched, it must encounter no surprise, it must have time to be carried out under favorable circumstances. This system of protection, covering the attacking troops, is the second duty of the advance guard. What does it do for that purpose? It holds the strong points which guard the main body against enemy attacks, it facilitates the action of that main body. In this case: Ruffey, Epirey, then Pouilly.

It is soon noticed that the enemy is undertaking nothing of importance towards Epirey. The battalion which occupies that village is then called back, two companies only being left: still employing a minimum.

The battalion marching on Pouilly makes important progress; the attack of that locality is going to begin. To insure its success it will be participated in by the battalion returning from Epirey. The attack of Pouilly is, of course, prepared by the artillery, the two batteries of the brigade, which advances to within good range of the place.

Here again we find complete application of the principle of the economy of forces.

To observe Pouilly, a start is made in that direction. The enemy is observed towards Varois and Saint-Apollinaire; protection is obtained by occupying Epirey. The fate of Epirey being settled, the attack of Pouilly is begun, but not till then. That provides, with 1 battalion to be moved from one point to another:

6 companies at Epirey if the enemy attacks;
8 companies at Pouilly when he is attacked.

Again, here is an example of the art of providing numbers, above all and before all, by maneuver; and this without abandoning anything, thanks to the occupation of strong points, the strength of which provides a means for the weak troops holding them of fulfilling again their mission, and of holding out.

In the same way, the attack is carried out as much as possible against one point and not against a whole front; and one point, moreover, which is generally a salient or a flank, because the attack can then make most favorable use of the numerical superiority which it possesses.

By acting thus, it envelops the adversary. On its enveloping line, more extended than the line enveloped, it can place more rifles and more guns than the enemy. Also, on that long enveloping line it finds the space and ground required for approach, for safe maneuver, for the final thrust of the mass.

It thus finds the means of employing with advantage the two arguments which it uses: fire and shock.

One of the points of the line of resistance being carried, the latter falls easily.

So, under the effort of the two battalions arriving behind a strong curtain of fire, able to maneuver and to use cover because they dispose of much ground, Pouilly is captured in spite of the fairly energetic resistance of the castle.

Behind Pouilly is the *chief line of resistance,* which includes the Fabrique, large square building with an interior courtyard, and La Filotte and Saint-Martin, both fortified, surrounded with trenches held by important forces.

The battalions which have carried Pouilly seek to debouch from it, together with the brigade's 2 batteries which have accompanied the attack up to about 400 meters from the fringe of the village. These troops are immediately stopped by French artillery which has reappeared to the east of the Langres road.

General Kettler moves forward 2 battalions of the main body, the advance guard being no longer sufficient. It has made an effort which has exhausted it, and a new enterprise is beginning: it requires some fresh troops.

As regards the units which have carried out the assault, they are reassembled, reorganized, joined to the main body. Part of them, however, first assure the occupation of Pouilly, starting point for a new attack. Thus will the attack be always carried on. Any progress made is definitely used by efficient occupation of the point carried, guarding it against counter-attack and insuring its permanent possession.

The attack progresses gradually, casting out anchors as it moves forward over a sea always full of surprises.

The attack of the Fabrique is then undertaken.

1 battalion of the 61st deploys between Val-Suzon and the road;

1 other battalion of the same regiment follows the valley itself. That one pushes back enemy groups between La Filotte and the Fabrique; it protects itself by means of 1 platoon against Fontaine.

Two companies are in the first line. The one on the left makes use of the railroad cutting. Both eventually reach the quarry, 200 yards to the northwest of the Fabrique. Further progress is made impossible by fire from the Fabrique and by fire from Saint-Martin.

The two companies then establish themselves facing Saint-Martin. Another company is called from the second line.

Again the Fabrique is attacked: another check, because the Fabrique is walled-in on every side and the artillery has been unable to prepare the attack: it is night, the attack fails with enormous losses; it leaves a flag in possession of the enemy.

Retreat is decided on, but the second battalion, which has carried out the attack through the valley and whose men are in some confusion, finds difficulty in disengaging itself.

In order to hold the enemy in place, the brigade is assembled south of Pouilly; it remains there till 8 P.M. It withdraws then to Vantoux and Asnieres. It had suffered an additional loss of 16 officers and 362 men.

The engagement had shown, by the evident occupation of Talant and Fontaine, and by the serious resistance of the enemy, that the French still had, on the evening of the 23rd, all their forces at Dijon and its neighborhood. The desired result was then obtained. On the following days, General Kettler maintains his troops opposite to, and within a short distance of, Dijon.

Many tactical lessons could be drawn from these battles around Dijon. As regards our main interest just now, the question of *protection,* they clearly show into what offensive tactics has developed the protective mission of the Kettler Brigade. We have seen to what extent its chief has carried out his duty.

The result, as we know, was the great success of the German Army of the South.

As to Garibaldi, those continuous attacks of January 21st and 23rd have convinced him that he is meeting important German forces. He has been content with a cautious defense. The result: disaster for the French Army of the East.

Error is human, you may say, and it is not a crime. But the crime lay in that Garibaldi, having been ordered to join the Army of the East, has not joined it. To carry out the order has never entered his mind. His behavior has been dictated by personal opinions and the desire for personal success.

If he had sought to obey, no material obstacle would have prevented it. The Pélissier Division left at Dijon sufficed to occupy General Kettler; the Army of the Vosges could safely join the Army of the East.

Garibaldi and General de Failly, two leaders of very different origin, cause the same result, disaster, by the same means: *lack of intellectual discipline, neglect of military duty.*

Obedience would have presented no difficulty in either case; but it had to be understood, and for that reason thought out; the spirit of discipline was necessary. To avoid the mistake, to prevent disaster, that was obtained by mere obedience.

But there is a lesson of a still more exalted kind to be drawn from these examples:

In our epoch which believes that ideals are unnecessary, that science and realism govern all things, we still find, to avoid mistakes and disasters, the sole observance of two moral principles: *duty* and *discipline*. And these two principles, in order to provide success, require, as shown by the example of General Kettler, *knowledge* and *reasoning*.

VI

THE ADVANCE GUARD

WHEN our flank guard brigade halted, on the evening of August 4th, at Woelfling to protect the army corps at Kalhausen, Achen, Etting, etc., it protected itself in turn by outposts. What was expected of these outposts? The possibility of resting in safety from enemy attack. It was necessary, therefore, that if the enemy appeared he be kept at a distance, and that if he attacked in force he be held long enough for the brigade to safely leave its now dangerous camp and take up battle positions.

To rest and to leave camp in safety means protection, that is material protection indispensable to rest and therefore to the condition of the troops, to preservation of their morale, of their confidence in the chiefs.

But what did the army corps expect from the brigade?

It expected it, during the night of the 4th-5th and during the day of the 5th to:

(1) Protect the army corps, at rest or on the march, against any blows, insure its material safety;

(2) If the enemy appear, hold him during such time and at such distance as will allow the army corps to continue its advance on Bitche, to act in accordance with orders.

The idea of protection, for which we only used one word, is therefore to be divided into two things;

(1) *Material safety*, permitting blows to be avoided when it is not desired, or not possible, to return them; that is the way to live in safety in the midst of danger, to camp or march without danger;

(2) *Tactical safety*, permitting a program, an order, to be carried out in spite of the obstacles created by war, in spite of the unknown, of the enemy; permitting safety of action whatever the enemy may do, by preserving one's own freedom of action.

Material surprise destroys material safety; it allows the enemy to shell at will our camps, our bivouacs or our columns on the march.

Tactical surprise destroys tactical safety, freedom of action. That would have been the case with the 5th Corps in 1870, with the Lespart Division in particular, if the enemy had appeared on the 5th or 6th of August. The troops on the march had to accept battle on the road they were following. Instead of continuing their movement, they must fight; they could not have arrived.

The same army corps was surprised materially and tactically at Beaumont on August 30th.

Materially because, in the absence of outposts, the enemy was able without difficulty to shell our troops in bivouacs first, engaged later in reassembling, that is unable to answer.

Tactically because in the absence of any system of protection at a distance, they were compelled to fight on the road they had to follow, and must renounce to proceed to the Meuse, which they had been ordered to cross.

That discussion shows what is understood by the word protection, aiming at action *sure* and *safe* through numerous and distinct ideas which must be first determined in each case, so that the service of protection resulting from them may duly fulfill its purpose.

From what we have seen, a certain number of corollaries can already be deducted:

(1) The instrument which insures the tactical safety of a large unit, of an army corps in the example taken, is the advance guard, by which general name we mean a detachment in advance, on the flank or in the rear, of the main body. Its position is immaterial, but its duty consists in employing what capacity of resistance it possesses for the benefit of the main body, to allow such main body to carry out the operation assigned to it, to comply with an order received. And as this operation or order vary constantly, we find that the advance guard's mode of action, the tactics it must adopt, are to be determined in every case by the nature of the operation to be undertaken, as well as by the conditions (time, space, ground, etc.) which govern the movements of the advance guard.

(2) The advance guard surrounds itself with its own service of protection, outposts or scouts, which, with some assistance from the main guard (occupying Singling in the example given) will suffice to insure the material safety of the unit protected.

(3) In every case, and whatever the conditions, we have seen, by the example of Rohrbach, that protection rests on two factors: *time* and *space*. It employs a third factor: the *capacity of resistance* of the troops.

The advance guard in that case was to guarantee to the advancing army corps, throughout the length of its advance, a radius of 3 kilometers at least inaccessible to the enemy for the whole length of the column, or rather for the time that the column passed a given point, four hours in the same instance.

But the commander of the advance guard also required *time* to reflect, to issue his orders, transmit them and have them carried out. He also needed *space* to move his troops before the enemy's arrival to such or such cross-roads and to deploy them there, providing for their retirement if they were swept back. One more problem of protection to be solved for the good working of the main guard.

In either case, if these conditions of time and space were assured by the protective troops, by the advance guard, the problem was solved.

(4) In any event, and chiefly as regards *space,* it is an essential principle of protection that a body of troops must always be master of the ground surrounding it up to the extreme range of modern arms, if it does not wish to be enveloped, surrounded, exposed to intense fire and destroyed before it is able to fight.

To this point, protection as we have seen it has only enabled us to avoid the enemy. The theory of war, however, allows of but one argument, battle; it is necessary to defeat the enemy, or nothing has been accomplished. Let us therefore consider the use of forces for the battle and in the battle. Shall we depend on protection? What shall we expect from it? To what extent and how can protection supply it? We are dealing with troops gathered for battle; they are no longer forces to be grouped, they are forces to be applied. The hammer is ready, we wish to strike. But the hammer is not yet

THE ADVANCE GUARD

in hand; ceaselessly our mass changes its form to advance, and up to the last moment it must keep moving.

At the same time, the point to be struck must be pointed out to the mass, and for that purpose we must recognize it. That point is not merely any point; sometimes it is a salient in the enemy's line, sometimes a flank, and in order to find it we must seek it first and reach it later.

The action of our forces would be quite uncertain, therefore, if it were undertaken without preliminary realization of two conditions:

(1) Determining the objective at which to strike;

(2) Handling and disposing properly the forces opposite such objective.

As long as this is not accomplished, we must retain entire freedom of action and protect the troops against enemy attempts; such troops would not be able to foil them efficiently.

On reaching the field of battle, at the moment of action, we are therefore obliged to solve the same constant problems of war: determining where and how to act, where and how to strike.

When that has been decided, we must still retain the means of striking even in presence of the enemy.

Protection is therefore still necessary, for it alone provides the possibility of avoiding surprise, of being safe, the possibility of seeing conditions clearly and of obtaining results in spite of the enemy, the possibility, in other words, of acting with safety.

Let us return to the instance of the 5th Corps, marching on August 6th.

With the dispositions we have suggested, it has been able to keep up a good rate of speed. The enemy has not appeared on the way; there has been no actual difficulty; but if the enemy had appeared he could not have prevented the movement or halted the column in its advance. In fact it arrives in good time on the field. With the formation adopted, the head appears at Reichshoffen at 9 o'clock, and the tail at 1 o'clock.

Are we going to thrust the forces into the raging battle gradually as they come up, drop by drop so to speak? Certainly not. It would be wasting them without serious results. Let us remember our principle of the economy of forces. We cannot win everywhere: it is sufficient that we win at one point. We must fight everywhere with a minimum of forces in order to be overwhelming at that point. We must know how to economize everywhere in order to spend lavishly at the point where we seek a decision; there we shall need *mass,* and we must therefore prepare a mass and keep it in reserve.

The 5th Corps, carefully economized, will therefore have to first mass, to then move, possessing necessary information, opposite the point of attack and to deploy there. It requires for that purpose to be duly protected, for one cannot mass, maneuver or deploy under the fire of hostile artillery any more than one can encamp under it. Elementary caution requires that blows be only accepted when they can be returned; art consists in then returning them more powerfully than they are received. The army corps deployed proceeds to attack; it will still need to protect itself on the exposed flank, for the same purpose of avoiding surprise which otherwise halts all attacks.

THE ADVANCE GUARD

Inasmuch as all the preparations for attack must be made under cover, we need protective troops. They will be known as the advance guard.

The advance guard, which has been necessary during the march to remove any obstacles from the route, is still necessary at the time of the engagement.

In Froeschwiller, we must also note that on its arrival the 5th Corps finds a condition well known, a position firmly held. The enemy has shown his forces and his intentions; he is held at certain points.

As we see, the advance guard of the 5th Corps, whose necessity has been proved, and whose use would still have been great, had to fulfill here only a part of the task which generally falls to the advance guard. It found the battle begun some time previously; at the moment of its appearance on the ground, the conditions were known, and the enemy located.

Quite different would have been on that same day the mission of an advance guard to the 1st Corps, intelligently organized. It would have had at the beginning of the day to

(1) Inform the commander of the actions of the adversary;

(2) Protect the moves by which the commander parried such actions.

Quite different again was the mission, two days earlier, of the Douai Division, advance guard for the Army of Alsace at Wissembourg. As such, it must:

(1) Reconnoiter the German forces entering Alsace;

(2) Protect the concentration of French troops destined to engage them.

Its tactics must therefore consist in:

(1) Attacking until the enemy showed forces greater than its own;

(2) Maneuvering afterwards by retirement on Reichshoffen, in order to hold out as long as might be necessary for the intended joining of the forces.

Once more we find that to bring the forces to the battlefield is not all; that protection which, by preserving our freedom of action, has allowed us to bring them there, must be continued by more protection allowing of their free disposal and of their employment wherever desired, whenever desired.

To that end, it must:

(1) Supply information as to the point or points to attack;

(2) Guarantee the possibility of bringing and deploying the main body opposite the chosen objectives;

(3) Protect the main body during all the preliminary operations.

Let us now take up the problem from a more general point of view: whether we deal with the 5th Corps arriving at Froeschwiller, the Army of Alsace in course of concentration, the Kettler Brigade reaching Changey or the Valmy Farm, the commander always has some maneuvering to plan, to prepare, to carry out.

The problem to be solved is no simple one:

It is in presence of the enemy that a decision must be taken;

It is with dispersed troops that we must act;

It is against an enemy free of his actions and capable of movement that we must plan and execute our maneuver.

THE ADVANCE GUARD

It is the duty of the protective troops to overcome these difficulties.

The *unknown* is a constant factor in war.

Everyone knows this, we may say, and inasmuch as everyone knows it, it will be duly feared, and cease to exist.

Such is not the case. All armies have lived and marched in the unknown.

It is from the "Sous-prefet" of Wissembourg that Marshal MacMahon learns of the Prussians' approach at the beginning of August, 1870. Until then he ignores their numbers, their points of assembly, the extent of their preparations.

But the 3rd German Army entering Alsace is no better informed.

On August 16th, 17th, and till noon of the 18th, similar ignorance exists at German General Headquarters. Armies badly commanded, we might say. But does not the same thing occur constantly in Napoleon's army? Remember the days preceding Jena, that day particularly which preceded the battle.

The best commanded armies have marched and maneuvered in the unknown, it was inevitable; but they have resisted that dangerous condition, they have come out victorious by depending on protection which has enabled them to exist without danger in an atmosphere full of peril.

It is in rear of the advance guard (5th Corps) of Marshal Lannes, accidentally running across the army of Hohenlohe on October 13th, that the whole French army will gather, and consequently find itself, on the 14th, in a good position for battle.

It is protection also which enables General Kettler to maintain himself with 5 weak battalions in front of the 40,000 men of Garibaldi, without fear of surprise.

To return to our theory, how shall that unavoidable unknown be overcome, how can we pierce the dense fog which always envelops the situation and the enemy's actions? By the advance guard.

Under the Empire it was at very short distances, in presence of an enemy clearly seen, whose power and position were readily ascertained, that dispositions were taken. Later, with increasing range and power of weapons, distances have increased, cover has been more eagerly sought, extended order more and more employed. By the smoke of his powder the enemy showed what positions he held. With smokeless powder conditions change, it is the unknown absolutely and continually. And to overcome the unknown which accompanies us to the very contact with our opponent, there is only one resource, the search till the last moment, even on the battlefield, of information.

And the information must concern the main body of the enemy.

At Pouilly, the Kettler Brigade meets bands of irregular troops which darken the horizon; it needs to see beyond them. An advance guard is sent. It disperses these bands, it undertakes the reconnaissance and attack of Pouilly.

Further back, it will find the main line of resistance of the enemy; its mission will be ended. In reality, with his patrols, with his detached troops of every kind, the enemy is everywhere. He is, however, with his main

THE ADVANCE GUARD

body at one point only, in one region. It is the main body we wish to strike, it is against the main body we must guard, it is about the main body that we need information. We must know where it is, and for that reason we must pierce the protective troops which evidently cover it. Our instrument of information must therefore be possessed of force. Yet that is not all: we must know what the main body consists of, and how much it is worth. The advance guard must therefore, in order to compel the main body of the opponent to show itself, force it to deploy; that requires attack, which means artillery and infantry.

To that extent must an advance guard carry out its mission of reconnaissance, and inversely that mission is carried out when such first duty, the obtaining of information as to the enemy's main body, is completed.

But there is another condition unfavorable to our maneuver, namely dispersion.

We arrive in column of route, or even in several columns: an army corps, 22 to 24 kilometers long, requires five or six hours to pass a given point, to enable its tail to join the head. During these five or six hours it disposes only of a part of its forces. The commander of the army corps cannot contemplate, however, pouring his troops drop by drop into even an action already directed. He must therefore assemble his troops, and then deploy them and place them facing their objective.

Under different circumstances, a different form of assembly must be assured.

The Army of Alsace in 1870 had to concentrate its 1st, 5th and 7th Army Corps before accepting any battle;

its advance guard, the Douai Division, could enable it to do so. The operation, which takes a long time, needs to be protected under penalty of being compromised. That is protection, and therefore the business of the advance guard. It must insure the arrival in place of all the battle troops in spite of the presence of the enemy.

To cover the assembly of the forces, later their employment, such is the second duty pertaining to the advance guard.

It implies especially time; resistance must be offered with weak forces over an extended front. How can that be done?

By strong positions, according to Carnot;

By maneuver when there is space enough.

The position which suffices for the desired result is, as we have seen, that which provides obstacles and a clear field of fire. Tactics will consist in hastening to such positions, in order to there begin the battle at long range by fire, to delay the decision.

It is the Kettler Brigade arriving before Talant and thrusting its advance guard companies on the two spurs north and south of the main road, and later at Daix; it protects, by that immediate precaution, the arrival of the main column and of the Turcey column (main body), holding a strong enough position to halt until further orders, with a small amount of troops (1 battalion, 2 batteries and 2 companies), the attack of far superior troops (several brigades of Garibaldi).

It is the advance guard of the Prussian Guard, on August 18th, hastening to Saint-Ail, then attacking Sainte-Marie-aux-Chênes. Seizure and occupation of the strong

THE ADVANCE GUARD

points necessary to the possession of the ground needed by the main body.

What is that ground, how much space is necessary?

We must evidently cover the place of assembly, but we must also cover the entry of the troops into action, cover therefore space enough for the deployment of the main body. The ground occupied by an advance guard may, for that reason, equal the length of the front.

A regiment acting as advance guard to a division can, in that manner, have to stretch over 1,500, 1,800, 2,000 meters or even more to cover the whole space required by the division. It can do so in safety through the use of ground under the conditions which we have examined.

That is not in contradiction with the limits prescribed by Regulations, with frontages determined there, because for the advance guard it is not a question of battle to defeat the enemy, and therefore not a question of battle formation.

The purpose is different: we propose to discover the enemy and to guard against him, over a certain space and during a certain time.

When we study the battle, we shall see what frontages to give to troops in order to obtain the result sought, the defeat of the enemy. To-day the result sought is different.

We must provide for the main body not only the ground necessary for maneuver, but also all the outlets which it may need for its deployment.

To insure possession, means to allow of carrying out in safety movement and deployment. The advance guard must consequently hold any positions controlling the outlet or controlling points from which the enemy might bring fire to bear on such outlets or on the place of deployment.

An example is provided by the advance guards of the 2nd German Army, establishing themselves, the 3rd of August, on the right bank of the Lauter, because on the following day the army is to cross the river: tactical protection.

Another example is that of the François Brigade of the Kamecke Division reaching Sarrebruck on August 6th. The bridges are intact, it immediately seizes them. In that way it controls, for its army corps, the means of crossing the Sarre. But that only insures the actual ground; it must still guarantee its free use, the right of crossing the bridges. That can only be done by seizing the heights on the left bank of the Sarre; they are a tactical necessity, and are accordingly seized.

The function of the advance guard is then not only one of protection, of material safety; it is also one of preparation, of tactical safety.

But as long as we have not beaten, or at least attacked, him, the enemy has his freedom of action, and can accordingly alter his dispositions or avoid the maneuver which we plan.

The first two results obtained by the advance guard, reconnaissance of the enemy and safety for the movement of our troops, would be of no use to us if the advance guard went no further.

Remember the morning of August 6th, 1870, at Woerth. From the Prussian camp the French are no longer seen, as they were seen on the previous evening. A reconnaissance is ordered for the purpose of ascertaining whether there has been any change since yesterday. The reconnaissance shows that the French are in force on the Sarre, but that they seem to be withdrawing; the sound of trains is heard from the Niederbronn station, and confirms the belief in a retreat. The advance guard attacks. It is most important that, at the moment battle is being prepared, the enemy should not be free to do as he pleases, to avoid the shock which is coming.

Reconnaissance will be followed, for that reason, by an attack of the advance guard destined to immobilize the opponent, especially if in the course of reconnaissance he has been found to maneuver.

The retirement of the 1st French Corps on August 5th would have allowed its junction before the battle with the 5th and 7th Corps; a junction which was only carried out much later, at the camp of Châlons, when two of these army corps were greatly reduced by battle, and the other had lost a great part of its morale.

The same happens at Spickeren, on the same day; General de François, after having carried the heights on the left bank of the Sarre, attacks the Rotherberg because he believes the enemy to be retreating. He must be held back, one does not strike an opponent who escapes the blows, one does not maneuver against an enemy on the move. He must first be immobilized; then only can the maneuver previously planned find its application; then only will that application be judicious, and duly respond to an effective disposition by the enemy.

These three conditions of war: the unknown, dispersion, freedom of the enemy, gave birth inevitably to the advance guard, and they determined the three duties which it must fulfill:

(1) Supplying information, and for that purpose reconnoitering until the moment when the main body attacks;

(2) Protecting the assembly of the main body and preparing its action;

(3) Immobilizing the opponent whom it is desired to attack.

These characteristics of the advance guard must never be forgotten in our dispositions. We shall give to each the share which belongs to it in accordance with the circumstances. It cannot be determined in advance, as it varies in every case.

The line of action naturally follows from this treble duty:

To reconnoiter, that is to see through the opposing protective troops, to reach their main body and force it to show itself, we must attack. We must also attack to conquer the ground necessary to our protective duties or our preparative duties. These are *offensive* missions which need to be carefully planned.

To hold the ground necessary for protection and for preparation when we have it and need only keep it, the action will be *defensive*. It is the case of the Kettler Brigade arrived and established on the heights to the east of the Changey Farm, and at Daix.

Under that form, advance guard tactics require capacity of resistance and power of endurance. They employ for that purpose everything that can develop these advantages: positions, strong points, clear fields of fire, rear guard actions.

THE ADVANCE GUARD

How can we fulfill simultaneously those three tasks which may require different means? How can we at the same time be on the offensive and on the defensive? It is evidently attack which we must employ, but keeping always the assurance of recurring efficiently to defensive if the necessity arises.

While we are conducting a systematic offensive with part of the forces, we shall keep the remainder for the occupation and organization of strong points in the rear; we shall take care, of course, to advance that line of resistance as the offensive progresses.

From this rough outline we can get an idea of the varied forms assumed by an advance guard engagement, according to the degree of progress and to the difficulties presented by each of the three questions to be solved, and according also to the special conditions of time, place, space, under which it is developed.

There can be no one form, no one type of advance guard engagement; there can be no one formula covering it always.

The same is true of the composition of the advance guard. It is determined by the treble duty to be fulfilled:

To reconnoiter, we naturally need cavalry; but we also need infantry and artillery to overcome the enemy's first resistance, to reach his main body and compel it to deploy;

To protect and to hold out, we need the means of fire at long range; troops capable of resistance, of firmly holding ground: infantry and artillery;

To immobilize the enemy, we must of course resort to the offensive, conduct it far enough to threaten the enemy with assault, or otherwise he can escape: we need infantry.

To such an advance guard of infantry, cavalry and artillery there has been opposed sometimes an advance guard of cavalry exclusively.

We have already seen how, even from the point of view of information, cavalry alone is insufficient. That insufficiency is more apparent still if we consider the two other duties of the advance guard.

Cavalry, shock troops par excellence and therefore suitable for immediate results, fulfills by no manner of means the condition of prolonged resistance which we seek.

Similarly, it assures only imperfectly space, because it does not *hold* the ground.

Reinforced by artillery, it offers a great power of action and of resistance, yet it is difficult for it to conquer and to keep.

Moreover, if beaten only partially, it risks very much to lose its artillery. The latter, in order to work efficiently, requires to be guarded closely and strongly; it needs the help of infantry.

The advance guard must therefore have each kind of arm, and as it acts independently it must also have one sole commander.

That theory of the advance guard which we have sought to establish by reasoning is a true characteristic of the tactics of Napoleon. I will prove it.

Napoleon wrote:

"An army must every day, every night and every hour be ready to oppose all the resistance of which it is capable. It necessitates that the army's various divisions be constantly in condition to support and protect themselves; that in camps, during halts and on the march, the troops be in such positions as are necessary to every battlefield, namely: the flanks to be strengthened, and all fighting units able to be employed under the most favorable conditions. To fulfill these conditions when in column of route, we require advance and flank guards to protect the front, the right and the left, far enough to enable the main body to deploy and to take position.

"The Austrian tacticians have always neglected these principles by formulating plans in accordance with uncertain reports, which reports, even if correct at the time the plans were drawn up, ceased to be so one or two days later, that is when operations had to be carried out."

From this quotation two necessities clearly appear for the Emperor:

(1) Necessity of constantly being able to oppose to the enemy all the power of which he is capable, and necessity of organizing forces to that end under all conditions: marches, camps, bivouacs;

(2) Necessity, at the very moment when an action is undertaken and a plan adopted, of doing so only in accordance with reports which at that time are certain and true; necessity therefore of organizing a service of information able to supply such reports.

How shall these two necessities be complied with?

"An army must be constantly ready to oppose *all* the resistance of which it is capable."

It is however impossible to keep all the forces of an army, even of an army corps, even of a division, in a state which permits the employment of all its resources when such forces are in camp, halted, or still more when they are on the march. One cannot camp, halt or march in battle formation. How can the necessity clearly stated by Napoleon be reconciled with the absolute impossibility, evident in practice, unless it be by the use of that property recognized and developed in the troops, capacity of resistance, which allows a certain portion, the advance guard specially organized and disposed for the purpose, to receive in any case the adversary and to stop him long enough for the main body to assume battle formation?

Moreover, every effort which troops can and must make to overcome the enemy need not occur at the same instant. The commander of an army corps, if he dispose of eight infantry regiments, will not engage them all simultaneously, even if he has them assembled at hand. Unity of time does not imply unity of instant. When we undertake the study of battle, of the action of force, we shall see the battle to be divided into a series of operations, into a succession of efforts aimed:

(1) Some at directing the command;

(2) Others at absorbing the enemy's activity, at wearing him out;

(3) Others, finally, at breaking by violent shock the equilibrium established between the pressure of the assailant and the resistance of the defender.

Sound employment of the forces, even if aimed at complete decision, entails the idea therefore of progress, of succession in the supply of forces. The first forces engaged must have produced their effect in order that the others may benefit by it. That idea of succession, once again, is not contrary to unity of time; it is necessary that time intervene to allow the troops and the armament to produce all the results of which they are capable.

Thus, to give only one example, in the attack of a locality we see:

The artillery: (1) Silence the enemy batteries which protect its approach;

(2) Prepare the attack by so sweeping the objective as to render it untenable;

(3) Accompany the attack.

The infantry: (1) Discover the ground held by the enemy;

(2) Besiege the position, sweep it with rifle or machine-gun fire;

(3) Assault it.

So many successive operations, alone permitting the best use of forces and requiring only at the last moment the simultaneous action of all these forces.

But the combination is then found.

From the fact that it is impossible to maintain all the troops constantly ready for battle, and that it is only necessary at first to keep a part of them in condition to be immediately used results the answer to the question.

Of the forces available we must always make, in camp, on the march and even at the opening of the battle, two divisions:

The *Main Body* and the *Advance Guard;* an advance guard sufficient to fulfill all the first requirements, and insuring for the main body the possibility of appearing at the proper time and place for the development of all the power which the troops can supply.

There is yet a second necessity stated by Napoleon, of making plans in accordance with reports that are certain and true at the moment the struggle begins . . . " and for that purpose," he adds, " we require advance and flank guards."

Independently of the duty already explained for the advance guard to maintain constantly the readiness for battle of the troops protected, there is then a second duty, the duty of obtaining information through:

Certain reports, which pierce for that purpose the apparent line of the enemy's service of protection and bear on his main body;

True reports, reliable to the last moment, which necessitates that the source of information, the advance guard, preserve contact with the enemy, once he has been met, *up to the last moment.*

From that double necessity of preserving always for the troops their ability to battle, of supplying certain and true information up to the last moment, we again find the advance guard with its duty to:

Provide information;
Afford protection;
Engage and hold the enemy.

All this means that the advance guard is needed up to the moment when the main body becomes engaged by deployment and first action on the enemy.

I insist on that point because one readily admits, in practice, the need of an advance guard in front of a column of route; one admits it less before an assembled body; one often rejects it before troops extended.

As though, because we are assembled or deployed, we

THE ADVANCE GUARD 169

had the right to fly from one surprise to another, to maneuver in danger, to come suddenly under the hail of murderous projectiles which infantry sends as far as 2,000 meters, artillery as far as 4,000 meters; under a fire so brutal that it demoralizes any formation (especially in mass), unless the danger has been previously foreseen by the advance guard, and unless that instrument has been maintained till the last moment.

As though we could reasonably draw up plans according to reports which are uncertain, or which even if true at the moment of deployment cease to be so at the time of approaching the enemy, that is a long time later, considering the great distance at which deployment must now be carried out. To deprive oneself of the advance guard before engaging the enemy is to interrupt contact at the moment it is most needed, it is to permit modification by the enemy of a situation which has caused us to take some dispositions which will no longer accord with real conditions.

We accordingly find again the necessity, before any troop not engaged, of an advance guard (whose form and use vary) constantly facing the enemy, in condition to deal with him if he appears, never losing sight of him, compelling him to display himself and his intentions; allowing also the commander of the main body to avoid battle if battle be not in accordance with his wishes, or to accept battle under favorable conditions, that is after reflection and suitable distribution of the troops in accordance with information obtained.

The commander of the main body must be assured of freedom of action and of the possibility of applying a plan in accordance with his intentions. And as to when that plan should be drawn up, Napoleon gives us his opinion in a criticism of Alvinzi:

"What should Alvinzi have done? He should have marched in one mass, drawn up dispositions for the attack of Joubert's Division only on the very morning that it was recognized. It is a principle that no detachment must be carried out (we now say: deployment) on the eve of an attack, because during the night conditions may change, either by withdrawal of the enemy or by the arrival of important reinforcements enabling him to assume the offensive and to completely counterbalance the dispositions prematurely taken."

VII

THE ADVANCE GUARD AT NACHOD

(*See Map No. 4*)

HOW is fulfilled, in a specified instance and under its special circumstances, the theoretical duty of the advance guard which we have already explained? The battle of Nachod will serve as an example for us.

On June 22nd, 1866, the General Headquarters of the 1st and 2nd Prussian Armies, in Goerlitz and in Neisse, received the order to penetrate into Bohemia and to complete their junction towards Gitschin.

To the 2nd Army belonged the 1st, 5th and 6th Corps, together with the Guard. The army commander took his dispositions for carrying out the entry into Bohemia on June 27th by three roads ending at Trautenau, Eipel and Nachod. On the 28th of June, the advance was to be continued across the Elba by Arnau and Gradlitz.

The 5th Corps (Steinmetz), on the left, was to proceed by the road of Nachod, which the 6th Corps would afterwards follow.

The 1st Corps, on the right, used the Trautenau road, while the Guard advanced, in the center, by the Eipel road, ready to support either wing, according to the circumstances.

On the afternoon of June 26th the 5th Corps had:

Its *main body* in bivouacs, west of Reinerz, astride the road leading to Nachod;

Its *advance guard* at Lewin, advance party at Gellenau;

Its *reserves* and baggage at and beyond Rückerts.

Everything was ready for the invasion ordered for the 27th.

According to the march graphic received from the army commander, the 5th Corps must, during that day of the 27th, reach Nachod with its main body and Wysokow with its advance guard.

Complying at once with the intentions of the Higher Command, the commander of the army corps ordered the advance guard to push its outposts during the evening of the 26th as far as the Mettau which marked the frontier. He had learnt that a hostile army corps had arrived at Opocno and stretched to the north of that city; that enemy columns were also assembling near Skalitz; that the defile of Nachod was weakly guarded.

In consequence of these dispositions, the commander of the advance guard, General Loewenfeld, arrived in the evening of the 26th at Mettau with the advance party. He found the bridges cut, the Customs House and its neighborhood weakly held by a few Austrian detachments; he easily overcame them.

He then decided to continue his reconnaissance, to march on Nachod with his leading troops and to occupy with his outposts the heights which overlook that locality west of the Mettau.

Meanwhile, the bridges destroyed on the Mettau at Schlaney were rebuilt and allowed the Prussian advance party to carry out the crossing of the river.

The Austrians had occupied Nachod with a small de-

THE ADVANCE GUARD AT NACHOD

tachment: 1 half-company of infantry, 2 squadrons of cuirassiers, 2 guns. These troops withdrew without serious resistance, but informed by telegraph General Benedeck, in Josephstadt, at 8.30 of the occurrence.

Shortly after, the Prussian advance party had occupied Nachod with two companies of chasseurs which placed outposts ahead, on the Skalitz road. It had also seized the heights north and south of the road (in line with Nachod) with two half-battalions of infantry, each protected by outposts.

The remainder of the advance party, 1 battalion, 2 squadrons and 1 battery, bivouacked somewhat back, on the road.

The main guard (4 battalions infantry, 2 companies chasseurs, 3 squadrons dragoons, 1 battery and 2 companies pioneers) had advanced to the Mettau, and established itself for the night south of the road, on the height of Schlaney, with 1 battalion at the bridge.

On June 24th the 6th Corps had received at its bivouacs of Koppernig the order to place at the disposal of General Steinmetz, commanding the 5th Army Corps, the 22nd Brigade of Infantry, 2 batteries and the 8th dragoons. This detachment, under the orders of General Hoffmann, actually only joined the 5th Corps during the day of the 28th, on the battlefield. It took no part in the battle of the 27th, except the 8th dragoons which, by two prolonged marches, reached, during the night 26th-27th the bivouac of Reinerz and formed with the 1st Uhlans and a horse battery of the 5th Corps a temporary brigade under General Wnück.

In the evening of June 26th, the 5th Corps is therefore distributed as follows:

Advance party at Nachod;
Main guard in line with Schlaney (1 battalion and 1 squadron at the bridge);
Main body of the army corps assembled at Reinerz;
Reserves in Rückerts; baggage and ammunition column further back;
Hoffmann detachment ready to join the army corps.

On the Austrian side, the Army of the North, after concentrating in Moravia under protection of Olmütz, has moved about the middle of June towards the position of Josephstadt-Miletin for the purpose of entering Bohemia. To cover this move, General Benedeck decides, during the night of the 26th-27th, to push on June 27th the 5th and 10th Corps towards the outlets at Nachod and at Trautenau.

As regards the 6th Corps, orders for the night of the 26th-27th are to "leave Opocno on the 27th for Skalitz, where it will occupy positions, pushing an advance guard to Nachod. . . .

"This disposition is intended to cover the concentration of the army in the neighborhood of Josephstadt, now in course of execution."

What was the situation of the 6th Corps at the time that order was to reach it? It was encamped on the 26th north of Opocno as follows:

Jonack Brigade, with 1 regiment of Uhlans at Krowitz and at Waly, covered
- at Ohnischow by { 1 battalion Chasseurs, 1 squadron }
- at Spie by { 1 battalion, 6 platoons }

THE ADVANCE GUARD AT NACHOD

Hertweck Brigade at Dobruschka and Perlitz, covered at	Prowoz and Domaschin
Rosenzweig Brigade, at	Bohuslavitz and Pohor
Waldstätten Brigade at	Mesritz and Gross-Rohenitz
Reserve artillery (5 batteries) at	Prepich and Ocelitz.

Each Austrian brigade was made up of 1 battalion chasseurs, 2 regiments infantry, 1 battery of 8 guns.

Headquarters were in Opocno.

During the evening of the 26th, the Jonack and Hertweck Brigades, barely encamped and insufficiently protected, were alarmed by detachments of Prussian cavalry appearing by the Giesshübel road. The brigades had to take up arms, and did not regain camp until late.

The respective disposition of the two army corps, 5th Prussian and 6th Austrian, on the ground shows more clearly than any words how war is understood on either side, and how it will be carried out.

On the Prussian side we see:

An army corps assembled, astride the road it intends to follow, its reserves in the rear on the same road, ready to act with all its means, its chief on the spot and in efficient command. That constitutes truly a *force* and a *will* combined. When Steinmetz has started his army corps, he will moreover hasten to the advance guard; he will be in Nachod at 8 A.M. of the 27th.

We also see an advance guard, already holding the

road far ahead, at the Mettau, insuring the tactical safety of the army corps and opening the road for it; the advance guard so thoroughly realizes its mission that on the very evening of the 26th it has hastened to Nachod.

Early on the 27th, a flank guard will be pushed to Giesshübel to cover the movement. Giesshübel is in Austrian territory; to have occupied it on the 26th would have exposed the offensive intended. But it will be occupied to cover the movement of the army corps once it has begun.

Such dispositions clearly show the sentiment of action which fills in the highest degree the commander of the army corps and the commander of the advance guard. They insure by the advance guard (for preparation) and by the flank guard (for protection) the possibility of carrying out the action which is to be undertaken with all the forces well in hand and in one direction.

The intention is to act with the whole at one point; that is rendered possible by protection; the decision will be obtained as the result of the economy of forces which has governed their distribution in the column.

On the Austrian side, we find an army corps extended over a front of more than 10 kilometers, which allows it to camp, to live and to march in comfort, a condition satisfactory enough as long as there is no enemy but answering poorly to the necessities of war. Moreover, it is divided into five distinct elements: 4 brigades and 1 artillery reserve.

So that if the enemy, primary objective of all war combinations, should reveal himself, the 6th Corps is incapable of action because of its dispersion. It would have to assemble first, but there is no time; no protective troops guarantee the two or three hours of quiet necessary to the front of more than 10 kilometers over which it has stretched.

An engagement would also be rendered impossible by the splitting-up of the force. Instead of one army corps responding to one will, there are 4 distinct brigades, evidently bound to operate individually, each for its own account. After the dispersion of forces comes the dispersion of efforts.

The commander of the army corps, moreover, is at Opocno, far from his troops, more comfortably established undoubtedly for working, for issuing his orders, but the instrument which is to execute them is far away. It either will not execute them, or it will execute them badly.

The commander only sees the subjective portion of his task: the means of preserving and of moving his army. He has completely lost sight of the purpose for which that army is intended: *the battle.* Nothing is prepared to undertake one and carry it through under favorable conditions. The notion of war, the idea of action, have disappeared, and there only remains the work of General Staffs, never sufficient in itself to cause victory.

In spite of all, in spite of this absolutely false conception of war, the Austrians were to meet at first on the 27th, particularly favorable circumstances. In war there are other things besides principles, there is time, there is space, there is distance, there is ground and there is luck which cannot be governed. The Austrians will finally be beaten, however. Principles cannot be discarded in vain; luck changes, and the spirit eventually resumes its rights over matters and chance.

Favorable circumstances: yes, for if we study the early hours of the 27th, the 6th Austrian Army Corps marching on a wide front by three or four roads, of which indeed the one on the extreme right was alone metalled, could rapidly attain its objective, the Nachod-Skalitz road. The length of the route was from 10 to 15 kilometers for the infantry brigades, and 18 kilometers for the artillery reserve.

By starting at 3 A.M., in accordance with the order given, the infantry could arrive at 7 o'clock, the artillery at 9 or 10 o'clock (because of the poor condition of the roads), if the enemy were not encountered.

There was a cavalry division to protect the march.

On the Prussian side, there were 18 or 20 kilometers to Nachod for the main body of the army corps, 22 for the reserves, all on a single road.

The Prussians have, it is true, an advance guard at Nachod, but until 11 o'clock or noon it will be thrown on its own resources.

As early as 7 o'clock, however, it may be engaged against the Austrian Army Corps, in which case the circumstances will be critical, both for that advance guard so long isolated and for the army corps threatened with the loss of its only outlet.

The very duty assigned to each of the two army corps was an advantage for the Austrians.

The commander of the 5th Prussian Corps could only have one purpose in view: to obtain access to the mountains, that is to seize and hold the keys to them,

THE ADVANCE GUARD AT NACHOD

deploying his corps in advance of the opening, on the plateau of Wysokow and of Wenzelsberg. That result, if the enemy sought to contest it, could only be obtained by an energetic offensive.

But the offensive was the hardest thing to organize, because of the nature of the ground.

The commander of the 6th Austrian Corps might understand his mission in two ways, because he was free to either take up a position near Skalitz, or, if the enemy appeared, to attack him with the utmost energy.

In both cases results were easily obtained. If he decided to take up positions near Skalitz, he would find near Kleny a ground with a clear field of fire, and he could reach that ground on the 27th without material difficulty, because of the distances, and without tactical difficulty on condition of taking the most elementary precautions, for the enemy needed the whole day to place on Nachod the equivalent of an army corps.

The position having been reached, he had time to occupy it and organize it during the evening of the 27th and morning of the 28th. But he must at least have conceived the idea and sought to realize it.

Should he, on the other hand, adopt the offensive, he had ground favorable to maneuver, allowing a whole army corps to be offered, free of its movements, able to fight against forces which were in echelons within a long defile, which they could only leave through one exit. Yet, in order to obtain results, it was necessary to foresee such offensive, to prepare and carry out its logical accomplishment.

A comparison of the hours at which, as the result of chance, the movements began, shows that in this respect also fortune was keeping a precious advantage for the Austrian commander.

For the Prussian Corps, we have seen that nothing is accomplished so long as it has not obtained strong possession of the plateau west of Nachod, a plateau marked by Wenzelsberg, Wysokow and the heights of Nachod. Any enemy occupying these points endangered the whole scheme. But these indispensable strong points must be sought 24 kilometers away.

If the Austrians adopted the idea of an offensive for the purpose of throwing the Prussians back into the defile, they must first seize Wenzelsberg and Wysokow and use them as strong points and starting points for all their movements to that end.

If, on the other hand, they followed the plan of taking up positions near Kleny while preserving the right to further action against the defile, it was still the possession of Wysokow and Wenzelsberg which they must obtain.

These points were not more than 13 kilometers from the two Jonack and Hertweck Brigades, which could moreover receive prompt assistance, and therefore spend themselves lavishly.

Thirteen kilometers as against twenty-four. If the Prussians and Austrians started at the same hour the Austrians could therefore fight during three or four hours with an undeniable superiority of numbers.

But in reality, the Prussian plans for the 27th were two hours behind those of the Austrians. At 5 o'clock in one case, at 3 o'clock in the other, the start was to be made. Altogether there would be an advantage of five or six hours in favor of the Austrian brigades. During all that time they would only face a weak advance guard of six-and-a-half battalions. They also had a cavalry division.

Such were the particularly difficult conditions (they could be partly foreseen, that is why I mention them) under which the Prussian general had to bring his column out of a long defile, and then deploy it.

To insure the success of these operations was the duty of his advance guard, a very heavy duty if we remember the numerical advantage which it must overcome, and the length of resistance it must provide.

We shall now see how the advance guard succeeded.

The Evening of June 26th in the Two Headquarters

On the Prussian side, we already know that the advance guard of General Loewenfeld had advanced on its own initiative from the Mettau to Nachod. That was a mistake as we shall see later, a risk which Steinmetz should not have allowed because it might endanger both the advance guard and the movement of the army corps, which was more important.

Steinmetz also issues his orders for the 27th. Those orders are the natural development of the scheme already outlined on the 26th. They state that:

"The army corps will march on Nachod and proceed past that point, in a westerly direction.

"All troops will leave camp or bivouacs at 5 A.M.; three ammunition columns will follow the reserves without interval; the other ammunition columns, together with the bridging sections, will proceed to west of Reinerz, where they will await further orders.

"Regimental baggage will be parked at Lewin; field hospitals at Lewin; heavy baggage west of Rückerts.

"The Hoffmann Detachment will send, on the evening of the 26th, the 8th Regiment of Dragoons to the main body of the army corps, where it will form, together with the horse battery of the artillery reserve and the 1st Uhlans, a brigade under command of General Wnück. With the remainder of his detachment, General Hoffmann will protect the left flank of the army corps, particularly in the direction of Giesshübel, and he will await further orders at Lewin."

On the Austrian side, General Raming has issued his orders in accordance with previous instructions from General Benedeck, sending the 6th Corps to Josephstadt, when at 1.30 A.M. he receives a new order dated from Josephstadt, 8 P.M., prescribing to march on Nachod.

That is because the Austrian General Headquarters has learnt the advance made by the 1st Prussian Corps during the 26th towards Trautenau, the approach of troops towards Braunau, the gathering of strong bodies at Reinerz and at Lewin, and their presumed march the next day towards Nachod. From this information it had been decided that the 2nd Prussian Army must intend to shortly enter Bohemia.

Benedeck persisted in his intention to concentrate on the Jaromer-Miletin position, without any intention to maneuver against the Prussians as they debouched from the mountains.

His 8 P.M. orders read:

"It appears from the latest reports which have reached me that strong enemy detachments are advancing on

Polié, Trautenau and Starkenbach. Consequently I give the following orders:

"The 6th Corps will leave Opocno on the 27th inst. at 3 A.M., and will proceed to occupy positions in Skalitz. An advance guard will be pushed on Nachod. The 1st Cavalry Division will be under the orders of the commander of this corps. The cavalry will be careful to explore at a distance, by means of strong patrols in advance, and on the flanks, of the columns.

"The 10th Corps will start to-morrow, 27th inst., at 8 A.M., after the early meal. It will leave its heavy baggage close to the fortress (Josephstadt) and occupy positions at Trautenau. An advance guard will be pushed forward. The 2nd Regiment of Dragoons will be attached to this corps. Detachments of cavalry will maintain communications between the 10th and 6th Corps, placed on the right of the 10th, and cover the left flank towards Arnau and Hohenelbe. The brigade detached at Praussnitz-Kaile will rejoin the corps at the time of its passing.

"The 8th Corps will proceed to-morrow to Tynist, in the neighborhood of Josephstadt, and will occupy the position abandoned by the 10th Corps. The 3rd Corps will leave to-morrow Königgrätz and establish itself on the left of the 4th Corps. A brigade will precede, and search the roads towards Jicin and Neu-Paka. The 2nd Corps and 2nd Division of Light Cavalry will proceed, on the 27th inst., from Senftenberg to Solnitz, to arrive on the 28th via Opocno at Josephstadt, and they will camp, according to previous instructions, respectively at Neu-Plas and at Jasena. The 2nd Division of Light Cavalry will relieve the posts established by the 1st Reserve Division of Cavalry in Opocno, Dochkabrus and Neustadt.

"The 4th Corps will keep its present position. It will detach one brigade between Arnau and Falgendorf (northeast of Neu-Paka) to guard the railroad. The 3rd and 4th Corps must protect the army's left flank by means of cavalry patrols sent at a distance. The 2nd Reserve Division of Cavalry will proceed, on the 28th, from Holitz to Josephstadt and camp on the heights of Smiritz on the right bank of the Elba. The 3rd Reserve Division of Cavalry will leave Wamberg on the 27th, reaching Hobenbruck the same day, and on the next day (28th) in line with Smiritz, on the left bank of the Elba where it will camp.

"*This plan is intended to cover the concentration movement, now being carried out, of the army near Josephstadt. It must not, however, prevent an energetic march on the enemy, if the opportunity occurs, but without pursuing him too far.*"

If we discussed this order, we should find in it the same characteristics as in the disposition of the 6th Corps on the evening of the 26th. There is a lack of military spirit. It is not an army striving to act with unity and with force against the enemy, it is a number of army corps, a quantity of divisions, moving like inanimate things, like pawns in the game, over a certain ground, without the directing motive of the Higher Command being shown anywhere, unless it be at the end of the order, to state the result not to be sought, not to be assured, but certain as a result of the order. "*This plan is intended to cover . . .*" But, if the order thus minutely detailed should be rendered impossible in some particular as the result of the interference of an enemy who cannot, after all, be forgotten in war, where he is the directing thread, what can still guide the various leaders? On the other hand, all their legs and arms have been carefully bound by the decision of means to be employed, by a list of childish recommendations. If the enemy appears, if the means prescribed should not fit the circumstances—and that always happens—they must either disobey or let themselves be beaten, solutions both of which lead to disaster.

The evil is visible to the Commander in Chief. He seeks to forestall it by adding at the end of his order: "It must not, however, prevent an energetic march on the enemy, if the opportunity occurs, but without pursuing him too far." He only increases such evil by creating doubt and confusion in the mind of men to whom he says simultaneously: "Withdraw and advance. Take up positions, and however march with energy on the enemy if the opportunity occurs." As if, in order to march with energy on the enemy, it were not necessary to seek him, and to plan accordingly. The worst consequences will result from this kind of command. It will be always thus when the Higher Command, lacking in broadness of view or in strength of will, seeks to substitute itself to its subordinates, to think and decide for them. In order to think and decide correctly it would need to see through their eyes, from the point where they stand; it would need to be everywhere at one time.

Command, in the sense necessitated by the scope of modern engagements, must mean for the Higher Command merely the clear determination of a result to be obtained, of the mission given to the subordinate unit in the operation undertaken by all the forces. But that determination must leave to the subordinate chief full liberty as regards the means to employ for obtaining the desired result in spite of circumstances which will arise to prevent it, which circumstances cannot be foreseen in advance.

Along these lines, the army commander, after communicating to the commander of the 6th Corps at Opocno *all* the information concerning the enemy which might interest him, after informing him of the movements of the army, would merely have ordered:

" For the purpose of protecting the concentration which is going on at Josephstadt, move to Skalitz, from where you will hold the Nachod and Kosteletz roads. You will dispose of the Reserve Division of Cavalry."

As a matter of fact, General Benedeck's order leaves Josephstadt at 8 P.M. At 8.50 news is received from Nachod, and it is several times confirmed during the night, that the enemy has carried and occupied that locality. This information, which would have been of the greatest interest for the commander of the 6th Corps, is not communicated to him. Raming will leave on the following day, ignoring the presence of important enemy forces within 12 kilometers of his leading troops. How could his dispositions have accorded with the reality?

In any case, he only received at 1.30 A.M., in Opocno, the General-in-Chief's order dated 8 o'clock, in spite of the short distance (15 to 17 kilometers) which that order had to cover. He immediately altered his preliminary dispositions, and ordered at 2.30 A.M.:

" The Hertweck Brigade will march via Bestwing,

Spie, Neustadt and Wrchowin on Wysokow, where it will face east;

"The Jonack Brigade will march via Spie, Neustadt, Wrchowin, Schonow and Prowodow on Kleny.

"The Rosenzweig Brigade will march on Bohuslawitz via Cerncic, Krein, Nahoran, Lhota and Spita towards Skalitz and take up a position north of the locality, on the right bank of the Aupa, facing the east.

"The Waldstätten Brigade will march via Rohenic, Slavetin, Rostock, Nauzin and Jessenitz towards Spita and Skalitz, where it will take up a position facing the east.

"The Hertweck Brigade will start at 3 A.M.; the Jonack Brigade at 3.30; the two others at 3 o'clock. Regimental baggage only will be taken. Stores will be sent to Opocno."

The artillery reserve was to proceed to Kilow, following the Waldstätten column, the sanitary service to Xajezd, the field hospital to Schweinschädel, the ammunition columns to Josephstadt.

The great distance which separated the 6th Corps Headquarters in Opocno from the spaces occupied by the brigades caused the order to be delivered late. Certain units even received it only after the time when its execution should have begun.

As we see, General Raming decides neither to take with his army corps a position from which to stop the enemy if he issues from the mountains, nor to throw him back by an offensive if the opportunity occurs.

He states neither of these results to be accomplished.

He even renders himself unable to attain either if, later on, he should come to a decision.

If the enemy does not interfere in the movement, he will finally at the end of the day have a situation full of weakness:

1 brigade at Wysokow;
1 brigade at Kleny;
1 on the right bank of the Aupa, north of Skalitz;
1 between Spila and Skalitz;
The artillery reserve at Kikow.

Each of these brigades is to take up positions on the ground assigned to it, facing the east: that disposition can only result in four successive and distinct brigade engagements, if we reckon the distances (7 kilometers from Wysokow to Skalitz, 4 from Wysokow to Kleny) and remember obstacles like the Aupa which separate the brigades.

The disposition forbids, in any case, the combined action of all the forces of the army corps at any time, either for maneuver and attack, or for resistance and counter-attack.

But if the enemy appears during the march to Skalitz, the same impossibilities appear: one cannot possibly oppose to him an army corps, but only four brigades and an artillery reserve, all acting independently, not to mention the other elements which are dispersed in various directions.

The complete lack of an objective still characterizes this order of General Raming, in which there is no mention of the enemy or of any tactical operation. How could the means be adapted to the end, and the forces properly directed?

The same order calls also for a few further remarks:

(1) The information received from General Headquarters showed the enemy as ready to attack, it mentioned him in particular towards Lewin and Reinerz. The alarm of the Jonack and Hertweck Brigades during the evening of the 26th confirmed that forecast. The right column of the Austrians, most exposed to attack, should have been well provided with cavalry and artillery. But it lacks cavalry; the cavalry regiment available continues to march with the Jonack Brigade, which has employed it until now for advance guard purposes. Lack of time probably prevents giving it sounder employment.

In the same way, the Hertweck Brigade disposes of one battery only, which is insufficient; the other brigades have one also, and they do not need it.

(2) The left column is followed by all the artillery reserve. It evidently does not need it. If it needs to be withdrawn and moved elsewhere, the movement will be a long one, entailing much delay.

(3) We find the characteristic faults of a fixed and symmetrical disposition of the forces. The threatened flank is always too weak, the flank that is not threatened is always too strong.

Moreover, the main body is not free to act where desired and as desired. There is no main body in reserve, because there are no protective troops covering it.

If the enemy therefore appears, one cannot *avoid* the engagement. Worse still, one cannot *direct* it, for the distribution of the forces in space means that they will all became engaged simultaneously. To direct the action, it would have been necessary to *create reserves,* to prepare a maneuver, to conceal it and to carry it out. The thing is impossible if the enemy attacks; the four brigades will almost immediately be engaged. It will not be possible to dispose of them.

(4) Even before the army corps comes into contact with the enemy, its dispositions make any maneuver difficult because of the length of its front (10 to 12 kilometers); any change of direction by the heads of columns is a very slow operation; a change of direction by the flank is impossible; there is no depth.

Let us now examine the problem for our own account; let us presume we are at Opocno, having received an order which repeats the information received and tells us, for the purpose of protecting the army's concentration in Josephstadt, to move the army corps to Skalitz where it must guard the Nachod and Kosteletz roads.

If we do not encounter the enemy on that march we must take up near Skalitz a position from which we can take action on all the dangerous roads. Such dispositions being once taken, if the enemy appear by Nachod, for example, the army corps must keep him from debouching out of the defile; if necessary it must throw him back into it by an energetic offensive. In view of that possibility, the army corps must assure to itself the possession of the plateau which commands the defile of Nachod. An advance guard will be pushed in that direction towards Wysokow. For similar reasons, another will be pushed north. To obtain information concerning the enemy's movements, to hold every issue by which he may appear: such are the tactics to be followed after our arrival.

THE ADVANCE GUARD AT NACHOD

If, however, during the march our army corps encounters the enemy debouching from Nachod, no hesitation is possible; we must attack him, throw him back into the defile, guaranteeing to ourselves at the same time a future occupation of Skalitz. The army corps will then proceed in a formation enabling it to seek and undertake battle with all its forces, instead of being drawn into it gradually.

In either case we will not pursue further than Nachod the retreating enemy.

If we should not succeed in throwing the enemy back into the defile, we must nevertheless seek to reach the position of Skalitz; we must maneuver with that purpose in view.

These elements of the problem being known, how shall we execute the move? We have four roads, are we to use them all?

A division marching by itself on a road has a length of about 15 kilometers. A second division, following the first, occupies hardly more than 9 kilometers.

If the two divisions march separately by two roads divided by the distance d we need, in order to assemble them at a point a on the first road, the necessary time for the last portion of the second column to arrive, that is to cover 15 kilometers $+ d$.

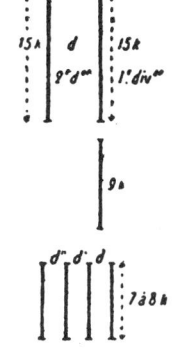

If the two divisions march one behind the other, the time necessary for assembly is the time necessary to cover $15 + 9 = 24$ kilometers.

In proportion then as $d > 9$, there is an advantage in taking one road or two.

If we divide the corps into four brigades over four roads, the time necessary for assembly is the time necessary to cover
$$8 + d + d' + d'';$$
and if therefore
$$d + d' + d'' > 24 - 8, \text{ or } 16,$$
it is a loss of time to use four roads.

But other considerations of a tactical nature appear, and limit more closely the choice of roads to follow.

While moving to Skalitz we wish to be prepared to meet the enemy, and to attack him even under favorable conditions if he appears. We need the power to maneuver up to the last moment, so we must be assembled. We must be able to change the distribution, to economize our forces, to advance them under cover. To meet these conditions we must organize in depth.

For these various reasons, we will only use two roads, that of Dobruschka-Neustadt-Wrchowin, and that of Pohor-Bohuslawitz-Cerncie-Nahoran-Lhota.

The average distance between them is only 4 or 5 kilometers.

The main body of the army corps will move by these two roads. In order to preserve its freedom of action and liberty of disposal of its troops, it must have a tactical advance guard. This advance guard must be on the side of the enemy, on the right, and be composed of troops able to *obtain information,* to *protect* the main body during a long enough time, to *hold* the enemy, that is to say we need:

Cavalry: we will use the regiment of which we dispose;

General Foch watching an attack in the Picardy sector.

Infantry: one brigade (the 1st);
Artillery: two groups.
Behind this advance guard, the main body will advance by the two roads.

On the right road: the remainder of the 1st Division, the corps artillery (so that it may promptly reinforce the advance guard);

On the left road: the remainder of the army corps: the Second Division, the Train, covered by an advance guard of material safety: a regiment of infantry, with some artillery and the divisional squadron.

On the other hand, the Prussian dispositions are not above criticism:

To move the advance guard to the frontier, a short day's march from Reinerz, was evidently not without danger, but it was justified by the general situation and also by the nature of the ground which compelled echelon formation.

But to push that advance guard any further, as far as Nachod, as General Loewenfeld did on the evening of the 26th, was to undertake an act of temerity full of serious danger. It was difficult to expect an advance guard composed of 6 battalions, 4 squadrons and 12 guns to resist five or six hours on the space necessary to the army corps' deployment, in advance of a defile, in presence of superior forces.

The presence of such superior forces in the neighborhood of Nachod was known. One knew an Austrian army corps to have been concentrated on the 26th at Opocno; one knew that other columns were assembling at Skalitz.

A violent attack should have been expected therefore for the 27th, when debouching, in the direction of Opocno or of Skalitz. One could not hope to debouch and deploy in the presence of such superior forces unless special dispositions were taken that would overcome the difficult conditions which would occur. These dispositions must consist in:

(1) Reducing the distance between the advance guard and main body, in order to support the advance guard as promptly as possible;

(2) Increasing the artillery of the advance guard;

(3) Marching in as solid and concentrated formation as possible, also for the purpose of shortening the crisis;

(4) Undertaking the march so as to arrive early on the 27th at Nachod. Instead of starting at 5 o'clock, starting at 3 o'clock.

To these prudent dispositions it can be objected that, as a matter of fact, the bold leap of the evening of the 26th made the Prussian Corps master of the bridge and crossing of the Mettau. Such an advantage was insufficient to compensate the dangers of the enterprise; for in view of the weak occupation of the bridge, the small importance of the river, things which were known, it was enough to push forward during the night the battalion of pioneers, to carry the enemy posts, to repair the bridge and establish additional crossings near the bridge so that the main body might carry out its march without suffering any delay.

But this risky action was also a very inconvenient one. The entry into Bohemia of the Second Army was to be undertaken at several points simultaneously and by surprise. That character of the operation was to be re-

THE ADVANCE GUARD AT NACHOD

spected in every army corps. The 5th Corps exposed the secret by moving, on the evening of the 26th, its advance guard on Nachod.

The Austrian command was to learn by telegraph, during the evening of the 26th, the news of the frontier's crossing by a Prussian advance guard; that gave it time to make counter-dispositions capable of preventing the operation planned.

In fact, as we know, the Austrian general-in-chief received at 8.50, in Josephstadt, the news that the post of Nachod had been attacked by much superior forces, and that it had consequently fallen back on Skalitz. It could not be expected that after receiving such news he would fail to take proper measures. At the time of giving to the Austrian commander such clear warning, one could not foresee on his part such lack of resolution.

A study of the campaign of 1806, which offers a similar case, would be most instructive. We should see Napoleon proceeding in a quite different manner to seize the crossing of the mountains, to prevent the enemy from defending promptly enough the points threatened. Thus, on the evening of the 25th, he would still have kept his forces in echelons, a fair distance from the frontier, giving thereby no indication as to the passages he proposed to employ.

On the 26th, he would have brought his army corps to the frontier after a long march; the head of the 5th Corps would have reached Schlaney, and the whole corps would have assembled solidly between Schlaney and Lewin, forming what the Emperor called a Mass of War.

On the 27th, the head of the army corps starting at 3 A.M. would have reached Nachod at 4 A.M. (from Schlaney to Nachod there are 3 kilometers). The tail of the army corps would have reached it two or three hours later.

What could the Austrian command have done then, even if active and decided?

On the 25th it can make no plan because the enemy's intentions are not shown.

On the evening of the 26th they become evident everywhere, and the Austrian command decides on measures for counteracting them.

On the 27th, these measures are carried out, but too late to bring the Austrian forces to Nachod before the enemy.

Napoleon tells us his methods himself, when he writes to Marshal Lannes, commanding the 5th Corps, one of the advance guard corps in October, 1806:

"On the 7th, you will camp between Hassfurt and Coburg (march of approach).

"On the 8th you will enter Coburg (or suppose Nachod) so as to reach it with your whole army corps, *and so that an hour before the arrival of your grenadiers there be no suspicion in Coburg of the beginning of hostilities.* Arrived in Coburg, you will take up a position in advance of that city, arranging so as to be on the 10th in Grafenthal, and you will prepare to support us."

And again, he writes to Marshal Soult:

"The Emperor orders that you prepare to enter Baireuth on the 8th, as early as possible. You will enter it *in mass,* so that an hour after the entrance of the first of your Hussars your whole army corps will be in Baireuth, and able to cover a few leagues more. . . ."

THE ADVANCE GUARD AT NACHOD

Development of the Battle from 3 o'clock till 8.30
(*See Map No. 5*)

On June 27th, 1866, the 5th Prussian Corps starts at 5 o'clock from Nachod. The advance guard, which was far ahead, starts at 6 o'clock only.

At 8 o'clock the advance party of 2 battalions infantry, 2 companies chausseurs, 2 squadrons dragoons and 1 battery reaches Branka, in front of Nachod.

It sends without delay as reconnaissances: on the road of Neustadt, 1 squadron dragoons and ½ company chasseurs; to the uneven ground north of Branka, towards Kramolna where enemy cavalry has been seen, it sends 1 company chasseurs; on the road of Skalitz it pushes 1 squadron and 1 battalion (incomplete through the delay of a ½ battalion) which are to occupy Wysokow.

Steinmetz arrives about 8 o'clock at Nachod; he is informed there that the advance guard has debouched without meeting the enemy. Expecting the day to go by without difficulties, he advises the 2nd Division of the Guard that he does not propose to use the assistance which it offered to him at Kronow.

The commander of the advance guard is busy issuing his outpost orders when the squadron pushed along the Neustadt road advises him, at 8.30, of the enemy's approach. When the squadron reached the plateau it had seen strong columns of all arms marching along the Neustadt road in the direction of Skalitz. They had already reached Schonow, Provodow and Domkow. The scouts, Prussian Dragoons, had met with a violent fire when approaching these columns.

The commander of the advance guard then orders the commander of the advance party to move to the Wenzelsberg plateau in order to hold the enemy, with all remaining available troops (1 battalion, ½ company chasseurs and 1 battery) to the cross-roads.

At the same time, he orders the main guard in Altstadt to advance towards the Wenzelsberg plateau along the Branka height which is south of the cross-roads.

Let us follow the advance guard in the fulfilment of its mission, and see how it carries out its three duties:

Reconnaissance: the Austrians being poorly protected, the Prussian Dragoons have easily noted the advance of two strong columns, one towards Domkow, the other towards Wysokow. That is more than required to make a decision. Reconnaissance is ended for some time.

Engaging the enemy, fixing him while one prepares to strike him; it cannot be done yet, the maneuver is far from ready. The army corps is on the way, its head will arrive only about noon.

Covering the assembly, and later the entry into action of the army corps is important and urgent; it is 8.30, the advance guard must suffice to that heavy duty during nearly four hours.

The task is first allotted to the 37th Prussians. That regiment has not fired a shot since 1815; it took no part in the minor Schleswig-Holstein conflict of 1864. It is the instruction of fifty years of peace which we shall see applied against the Austrian army which has fought recently (in 1859). We shall soon recognize, on one side men who know war without having waged it, the

THE ADVANCE GUARD AT NACHOD

Prussians; on the other side men who do not understand it though they have made it.

To the first enemy menace the Prussian commander replies by the following distribution:

1 squadron, ½ battalion and ½ company of chasseurs on the Skalitz-Wysokow road;

½ battalion moving towards the same point;

1 squadron and ½ company chasseurs on the Neustadt road;

1 battalion and 1 battery on the march to Wenzelsberg;

1 company chasseurs at Kramolna.

The remainder of the advance guard (3 squadrons, 3½ battalions, 2 companies and 1 battery) arrives from Altstadt on the plateau.

Very sound dispositions.

Everybody (less ½ battalion kept at Altstadt) is sent on the plateau because one must at any price cover the opening of the defile and prepare the entry into action of the army corps.

To cover it is necessary to hold the points from which the enemy might sweep the opening, as well as all the ground necessary to future employment of the army corps, as much in width as in depth.

And the advance guard immediately extends to a front of 4 kilometers; local circumstances make it necessary.

The danger thus created for the advance guard may be very great. That is of little importance, if it only *holds out* long enough for the arrival of the army corps, say till noon.

That is the sole idea of the commander of the advance guard, and from it will spring the form of engagement which he adopts; it is in tactics, and necessarily in tactics of details (because of the dispersion) that he will find the possibility of holding out in the present situation. We shall see how, on the other side, the Hertweck Brigade, actually advance guard for the Austrian corps, will fail to realize its duty as advance guard, and will simply begin a brigade engagement purely theoretical, poorly planned in the beginning and illogical in its execution.

Let us return to the Austrians.

The Hertweck column, having started at 3.30, crosses twice, in Spie and in Wrchowin, the Jonack column, which delays its advance. It enters into contact with the enemy at 7.30 through its advance guard (25th Battalion and 2 guns). At that moment:

The Jonack column advances on Domkow;

The Rosenzweig column marches on Lhota;

The Waldstätten column marches on Skalitz, where it is to assemble.

Meanwhile General Raming, having come to Skalitz and later to Kleny, finds there the commander of the 1st Cavalry Division who states that his outposts have been thrown out of Wysokow. He consequently issues orders that:

The Hertweck Brigade will continue its march on Wysokow;

The Jonack Brigade and Rosenzweig Brigade will move on Kleny, pushing 1 battalion on Wysokow;

The corps artillery and Waldstätten Brigade will march on Skalitz, in reserve.

The cavalry division will send its Solms Brigade to Kleny, and its Schindlocher Brigade to Dolau (5 or 6 kilometers from Kleny).

It was, therefore, the second interpretation of General

THE ADVANCE GUARD AT NACHOD

Benedeck's order which General Raming adopted, for, by all his dispositions, he was bringing his troops nearer to the defile.

But on adopting the offensive, he should have assured himself the possibility of:

(1) Taking it towards Nachod and Wenzelsberg;·

(2) Taking it safely and promptly, for he might have business in other directions.

However, his dispositions, imperfect as they were, increased the danger to the Prussian advance guard, for they meant that

The Advance Party

2 battalions ⎫ would   ⎧ 7 battalions ⎫ Hertweck
1 battery   ⎬ have to   ⎨ 1 battery of ⎬ Brigade
1 squadron   ⎭ encounter   ⎩ 8 guns   ⎭

and the   ⎫ 7 battalions ⎫ would have to fight
whole advance   ⎬ 2 batteries ⎬ against 2 or even 3
guard   ⎭ 4 squadrons ⎭ Austrian brigades.

Before seeing how the advance guard acts, let us examine the ground.

Position of the Plateau of Wenzelsberg

The Neustadt road follows a cutting, Branka-Schlucht then the western edge of the forest of Branka, approximately on the crest of the plateau. This forest ends on the east in steep and rugged slopes which extend to the Mettau.

The crest of the plateau rises with the road, passes later east of the road which still provides, however, with the edge of the forest a powerful line of resistance.

There is found the last strong point for troops which fight facing west, without possibility however of employing much cavalry or artillery, because of the woods and steep slopes which extend to the Mettau.

The plateau of Wenzelsberg dominates all the ground stretching west, especially the heights of Kleny. But sight is interfered with by covered ground, both in the Prowodow and Schonow direction, and in that of Neustadt and Wrchcwin. Assaulting troops coming from these directions find in the bushes and cuttings protection against fire and against sight of the defenders; such protection was doubly great on the day of the battle, because of the advanced condition of the harvest.

Thus, the line of defense of the plateau of Wenzelsberg facing west is not on the crest but halfway down the slope. It can be traced by the western edge of Wysokow, the southwestern edge of the small wood (Wäldchen), the Eglise Evangélique of Wenzelsberg, the Maison Fortière and groups of trees that surround it on the east and on the south as far as Sochors.

From that line runs a slope with an inclination of 5 to 10 degrees, entirely open, allowing powerful artillery and infantry fire. Moreover, the line of defense is high enough to dominate the heights of Kleny.

Its length is about 4,000 meters. It provides a good defensive position for an army corps. The distribution of the forces would then include:

The occupation of the localities by infantry supported by a few batteries; the massing of the cavalry and artillery between Wysokow and the Wäldchen wood to the south.

The massing of all the infantry reserve on the plateau

east of that wood; a detachment of infantry with artillery to be moved to the plateau, north of Wysokow.

The disadvantage of the position is evidently its lack of depth, and consequently the difficulty of moving, in the rear, any artillery or cavalry.

It is easy to see, however, that if it be occupied as I have suggested it would be difficult to envelop by its left, on account of the wooded nature of the ground. Attacked on its right by troops pointed at Nachod, there would be time enough, thanks to the detachment holding the plateau north of Wysokow, to take counter-dispositions capable of stopping the enveloping movement which could only be slowly carried out.

On the other hand, in front of the position are found Prowodow and Schonow, behind which the enemy can assemble and dispose his troops, and in case of necessity employ them as points on which to retire.

The stream which runs from Wysokow to northwest of Prowodow is absolutely negligible; it has numerous crossings, and is nowhere an obstacle; the meadows along its banks were dry and firm.

These particularities of the position became especially evident when the Austrians, thrust back on Schonow and Prowodow sought, in a violent effort, to envelop the right flank of the position; it was too late.

Considering the nature of the ground, what use would be made of it, and what could be expected of it on the 27th?

The Austrians, because of their position and overwhelming superiority, could and should evidently seize as quickly as possible the line of defense mentioned previously, and in that case the fate of the 5th Prussian Corps was already half sealed: Its existence became at least very difficult, for the Austrians, starting from these points, had every facility for ejecting any Prussians who reached the Neustadt road. The objectives to aim at were, therefore, the Branka wood and Branka height. The Prussians, on the other hand, must seek to occupy the position of Wenzelsberg and woods surrounding it, organizing there a first line of resistance, the second being formed by the Branka cut and western edge of the Branka wood.

As to the resources which that second line would offer, there are, however, some remarks to be made:

(1) On the Branka-Schlucht and Branka Wood front, infantry only could be employed; the ground on the east was not practical for artillery, and sight was very limited.

On the left, artillery would meet with difficulty in moving or retiring.

On the right, south of Wysokow, there were a few artillery positions, but it would soon have been exposed to fire from the little Wäldchen wood.

(2) On this front effective fire could very easily be had up to 500 meters.

There was little cover: the ditches of the road were insignificant, the woods cut, only a few large ones offering any cover.

(3) Retirement from the position was difficult because of the steepness of the slopes and thickness of the undergrowth.

Retreat might be cut from Nachod by an enemy marching on Altstadt.

On the other hand, behind this position was found

Hill 1113, which the Austrians might find profitable to occupy, because from there they swept the whole Altstadt-Nachod road.

Having superficially studied these two principal positions, let us examine more carefully the strong points which constitute the former, so that we may later understand what occurred there.

The wood north of Wenzelsberg (*Wäldchen* on the German maps) has a northeasterly edge of 300 meters, with a depth of 1,300 to 1,400 meters. It consists of a number of groups of young trees, of undergrowth and of clearings: a mixture destined to mix up the battle in the same manner. It is framed on its northeastern and southeastern edges by a sort of embankment.

The inside contains some ravines, the beds of dried-up torrents. The most important of these ravines leads near the northern edge. The battle was destined to reveal all its tactical importance. It has, almost from its beginning, a depth of 10 to 15 meters, about the same width, very steep banks. It is an obstacle to the movement of infantry. That part of the ravine which ends at the southern edge of the wood offers good cover for infantry, which finds there an excellent field of fire on the opposite slope, facing the Skalitz road.

The southwestern border is not clearly defined.

Wenzelsberg possesses, at its two northeastern and southwestern extremities, two solid buildings, the Evangelical Chapel and the church.

The Evangelical Chapel offers a very good strong point facing Meierhof, with a good *glacis* opposite that locality, which it dominates by 50 meters. It can easily be fortified with the entrenching tools. On the other hand, two hollow paths run parallel to it and provide an excellent approach for assault. Orchards run from there right and left, along Wenzelsberg, a small village of low houses, especially in its southern part. In the northern portion, the houses are more numerous, better grouped, more important, there is a better view from the gardens.

The Wenzelsberg Church is a solid construction, without any opening towards Branka but with an opening towards Wenzelsberg, a few windows right and left. To provide good protection it required a few improvements in the church itself and in the surrounding wall.

Between Wenzelsberg and the Maison Forestière we find an important ravine which is an impassable obstacle to artillery, cavalry or infantry under fire. At its two extremities it is more easily crossed. It provides an opening and means of approach to small detachments.

The Maison Forestière is a solid construction with annexes, having no surrounding wall, inappropriate to defense, easy to assault because of the cover which the assailant finds up to quite near the house.

The little triangular wood east of the Evangelical Chapel has no tactical value. It interferes with sight.

Against an enemy coming from Branka the Wäldchen or the church of Wenzelsberg, the surrounding woods provide good strong points; but the efficacy of the fire which they permit is limited, the advance of the enemy being partly concealed from the defense by the heights and woods of Branka. On the other hand, any debouching of the enemy from these heights and woods is always problematical. He cannot support his attack with artillery and cavalry.

Against an enemy coming from the southwest of Prowodow, or from Schonow, the localities held by the defense allow good use of fire; they constitute a dominating position, but their advantages are compensated by lack of continuity in the borders, by the covered paths and means of approach which the enemy finds in the little woods and ravines. The best strong point of the defense is the Evangelical Chapel.

BATTLE OF THE PRUSSIAN ADVANCE PARTY AGAINST THE HERTWECK BRIGADE. ENTRY INTO ACTION OF THE PRUSSIAN MAIN GUARD, 8.30 TO 10.30

General Hertweck, in Wrchowin, received from his Uhlans the first intimation of the presence of Prussian troops on the plateau of Wenzelsberg. He concluded from this information that the enemy must be established on that plateau, facing south. He decided therefore to meet him with 1 battalion and 1 company of infantry while with the remainder of his forces he continued to advance on Schonow, and thence against the enemy's flank.

This detachment, formed on the Neustadt road, easily drove back into cover the Prussian Dragoons. Between 8 o'clock and 8.30, the brigade was at Schonow.

The advance party, including the 25th Battalion Chasseurs and 2 guns, then moved from Meierhof towards the Evangelical Chapel of Wenzelsberg, occupying the chapel's courtyard with 2 divisions, extending 1 division along the orchards and gardens in front, and placing its 2 guns south of the church, from where they immediately opened fire on the advance party of the Prussian Guard.

Meanwhile, the brigade was massing northeast of Meierhof in battle formation, in three lines. The battery took up positions on the right of the front line, and immediately opened fire against the Prussian artillery.

The battalion previously detached on the Neustadt road had approached and occupied a small wood on the right of the battery, near the Maison Forestière.

About 8.30, the deployment being accomplished, a short rest was ordered, and between 8.45 and 9 o'clock the attack was begun.

On the Prussian side, we see at about the same time 1 battalion, ½ company chasseurs and 1 battery hasten along the Neustadt road, then take their direction on Wenzelsberg, and deploy the 2 half-battalions in the space between Wäldchen and the wood east of Wenzelsberg.

The battery took up positions between the two half-battalions. To protect the flanks one had also sent the ½ company Chasseurs into the Wäldchen, and scouts into the wood east of Wenzelsberg.

On the Neustadt road, the 3 squadrons of dragoons hastily brought from the main guard have come to support the squadron thrown back. As the Austrians had taken possession of the covered ground halfway up the slope, the cavalry could not act to good advantage on the left of the Prussian attack: it moves behind the right flank, in a good position for support.

Shortly before 9 o'clock, the Austrian brigade advances on the slope, both sides of the Evangelical Chapel, preceded by a line of scouts; it meets with a well-directed fire from the Prussian snipers under fair cover in the wheat, the woods and a small height east of Wenzelsberg. The brigade is compelled to stop, and to take cover near the Evangelical Chapel. At the same time, the Prussian battery has inflicted such important casualties to the right flank of the Austrian Brigade that the latter is compelled to withdraw on Schonow, 1,500 meters to the rear; the 2 Austrian guns near the Evangelical Chapel are also compelled by the fire of the Prussian infantry to withdraw on Meierhof.

At 9.15 General Hertweck orders to move forward the battalion forming the second line, and to let it prolong the right flank of the front line; the former third line is to draw closer to the front line.

The attack is then carried out: on the left by 2 battalions which, under cover of the gardens east of the village and of the village streets, advance on the Wenzelsberg church; on the right, 2 battalions march on the wood east of Wenzelsberg, supported by the battalion at the Maison Forestière which now moves north towards the same wood.

The Prussian battalion receives the attack of the enemy brigade at 500 meters, by well-directed and well-controlled fire. Fire is in short bursts from the supports running up to the line. At the same time, rapid and flanking fire comes from the Wäldchen and neighboring woods.

The Austrian columns stop, hesitate, start forward again; but then the Prussian half-battalions advance also, and execute at short range such a powerful burst of fire that the Austrian masses, completely demoralized, turn tail.

What has really happened? The 2 battalions on the

left flank of the Austrian front line have reached without any great difficulty the courtyard of the Wenzelsberg Church, and the northern edge of Wenzelsberg village; maintaining part of their forces on these points they try with the remainder several attacks against the Wäldchen. But they are repulsed by the fire of the Prussian Chasseurs who have just received reinforcements. Some of the units suffer heavy casualties in these useless efforts.

The two battalions on the Austrian right flank have been halted by some enemy forces, but especially by the wide ravine they must cross. Much disorganized by the passage of this obstacle, they have nevertheless reached to within 120 paces of the wood east of Wenzelsberg, where they have been completely stopped by the Prussian artillery's powerful fire. In disorder, they seek safety in the ravine.

While the 1st Austrian line was greeted in this fashion, the 3rd Gorizutti Battalion which had followed it in echelon formation, comes up to attack. It debouches near the Maison Forestière and crosses the upper portion of the ravine. The ½ battalion of the Prussian left flank faces left, and attacks in and between the clumps of trees from which the Austrians are coming, at the same time as 2 half-battalions of the Prussian main guard extend from the Neustadt road towards the Bracez cut, and rush to support the counter-attack already undertaken.

The Austrian battalion is thrown back, and slowly withdraws under fire towards Schonow. It is followed in that direction, and as far as Sochors, by 1 half-battalion of the 58th.

The other half-battalion of the Prussian main guard

occupies the clumps of trees east of the Maison Forestière, and the half-battalion of the advance party resumes its position by the Schimonski half-battalion.

Then occurs a pause in the battle, during which the Austrians are content to keep up fire against Wenzelsberg and its church. The battery of the Jonack Brigade, west of Prowodow, joins in, and General Hertweck draws up his second line for a new attack.

On the Prussian side, the 2nd battery of the advance guard has joined the 1st. Gradually, the battalions of the advance guard have reached the plateau.

The 2nd Austrian line's attack has occurred when 2 half-battalions of the Prussian main guard have come to support in the Wäldchen the half-company chasseurs which was already there, and which had already reached the edge while the other 2 half-battalions, as we have seen, reach the clumps of trees and the Maison Forestière.

The two battalions of the Austrian right hastily retreat into the ravine, from where they keep up a battle by fire which remains undecided, while the seizure of Wenzelsberg on the left may be full of consequences.

The 3rd Kellner Battalion and a few companies of chasseurs which have advanced as far as the ravine appear to have debouched on the Maison Forestière.

The position of the Austrians in the ravine and around the Maison Forestière was taken in the rear by the Prussians of Sochors and was already very critical when 2 more half-battalions reach Sochors. One Prussian half-battalion seizes the Maison Forestière, and the other advances in the wood southeast of Sochors.

The Austrian second line can no longer remain in the bed of the stream. It falls back on Schonow, where

212 THE PRINCIPLES OF WAR

General Hertweck assembles his troops. Alone, the 25th Battalion and half the 2nd Kellner continue to occupy Wenzelsberg and Wenzelsberg Church.

At the same time as these attacks are carried out, the Prussian advance guard has recalled all unnecessary detachments:

(1) The company of chasseurs sent to Kramolna has rejoined at Wysokow the ½ battalion already there;

(2) The Bojan half-battalion of the 3rd of the 37th, which had supplied the outposts at Nachod, arrives late for that reason; it is placed in reserve on the edge of the Branka wood, north of Wäldchen;

(3) The main body's 2 companies of chasseurs are in the direction of Bracez, holding the Neustadt road.

The distribution of Prussian forces at that moment is therefore:

At Wysokow: ½ battalion (Kurowski) with 1 company chasseurs.

Before Wenzelsberg 2nd of the 37th, flanked
{
In the Wäldchen by
  ½ company Chasseurs
  ½ of the 1st of the 58th
  ½ of the 1st of the 37th
To the south by
  ½ of the 1st of the 37th
  ½ of the 1st of the 58th
  2nd of the 58th
  ½ F   of the 58th
}
holding the Maison Forestière and surroundings.

Two companies chasseurs hold the woods on the Neustadt road; the ½ of F of the 58th remains in Alstadt.

In reserve, behind Wäldchen, are the Bojan half-battalion and the cavalry. At the southeastern corner of Wäldchen the 2 batteries of the Prussian advance guard fight against the Austrian artillery, west of Schonow and Prowodow, and against the masses of infantry which are seen in these localities.

The cavalry of the advance guard (4th Dragoons), finding no opportunity to intervene, remains under Austrian shell fire between Wäldchen and the Neustadt road.

That first act of the day calls for a few remarks. The very brilliant results of the Prussian tactics are undoubtedly due to technical superiority of fire and to a wise use of the troops. We find 1 battalion, 1 battery and 4 squadrons resisting 1 brigade (7 battalions) and 8 guns.

What course has the battalion followed?

It would have been simple and logical to meet the superiority of the attack by the cover and resistance to be obtained from the ground, by the occupation of Wenzelsberg Church, Wenzelsberg Village, and some of the most suitable houses. These cautious tactics would have had the disadvantage of breaking up the troops, of making their control, particularly as regards fire, very difficult. It would have given less results than the open ground which allowed the whole available power of fire to be used. The enemy is therefore met in line, in the open; but as in that position one might be turned, protection is necessary; on both flanks the woods are occupied.

Much help is given by the state of the crops, which conceal the strength and position of the Prussian troops.

In the same way, when the enemy has been repulsed he is not pursued. On the covered ground, pursuit would have disorganized the troops, and enabled the enemy to reopen the question of the results obtained.

I must also point out how the bold tactics of the 2nd Battalion of the 37th have been facilitated by the troops arriving singly, on its right in the Wäldchen and on its left towards Bracez, later in the woods around Sochors. All the commanders of these little units, poorly informed and without orders, show themselves capable of initiative by which they successfully interfere in the action.

The battle front at the end of this first act of the struggle is considerable for the troops available (1 brigade, 1 battery, 1 regiment of cavalry).

There are 2,500 meters from Sochors to the southern edge of Wäldchen;

There are 3,700 meters more from Sochors to the southern edge of Wysokow.

Moreover, the half-battalions are intermixed, the regiments no longer exist; barely can the half-battalions be reorganized on the ground. There is for sole reserve 1 half-battalion.

The circumstances explain and justify nevertheless the dispositions taken.

If all the ground be not held, the army corps on its arrival will not have the space necessary to its deployment. Moreover, by this line of action the enemy will be deceived as to the strength of the forces opposed to him.

The artillery enters into action from the very start. The 1st Battery opens fire with the battalion; when it finds itself endangered by the Austrian infantry fire, it withdraws to take up a new position where it is joined by the 2nd Battery of the advance guard. The two batteries then keep up a ceaseless fire, increasing thus the power of the infantry.

Being unable to act, the cavalry keeps close at hand, a constant threat.

The Use of Fire

The first attempt of the Austrians was stopped by the fire of the front line, assisted by its supports.

The attack is renewed. Four battalions make a frontal attack while one battalion attacks the flank. The Prussian battalion does not allow itself to be worried by this number. It is in perfect accordance with regulations, by individual fire and short bursts, that the line holds, and by rapid fire in the case of the protective troops.

The enemy continuing his advance, the battalion marches towards him and opens, at 150 meters, controlled fire the effect of which is evident. The danger has not disappeared altogether as the result of this brilliant success. An enemy battalion appears on the flank of the Prussians. The left half-battalion faces left, in good order and close formation, assumes the offensive and thrusts back the Austrians.

As we see, the Prussian battalion has held its fire and used it successfully; it has sacrificed all other considerations, such as the occupation of the strong point of Wenzelsberg, cover, pursuit of the enemy, preferring to give to its fire all possible effect, keeping for that purpose and till the last minute all the troops assembled and well in hand. These tactics of a Prussian battalion accepting in the open the attack of a brigade would not be sound today, when, in presence of a serious enemy properly armed, it would be necessary to hastily occupy the strong points or take cover. But now as then, the distribution of the troops would be affected by the need of keeping efficient control of fire.

Moreover, the Prussian battalion has known how to take advantage of:

(1) The weakness of the Austrian formations which it saw;

(2) Cover offered by the harvests;

(3) Superiority of armament, of course, and the best use to be made of it.

(4) Absolute fire discipline which rendered possible, up to the last moment, efficient control of fire by the commander.

Here are clearly seen the principles and aims of the Prussian school as regards the use of fire. Fire has become a power of the first order; it must be properly used, and for that reason remain under the *control* of the commander who has been instructed in its use. Superiority of fire will therefore depend, naturally, on armament, but it will depend still more on the use made of it, which necessitates men trained in shooting under war conditions, remaining constantly under the orders of an experienced leader.

The above example shows what I mean by *experienced:* I mean a leader schooled in the use of fire, its effects and the ways of obtaining such effects; a leader who therefore knows when his fire should be individual or by volleys, rapid or deliberate, who also knows by experience in what measure and within what time his men can obtain the result desired, keeping their self-control and discipline. He also knows when nervousness and physical fatigue will become apparent, how the men can then be taken in hand again, who realizes moreover that fire alone cannot give a decision, who combines it therefore with the advance by sound discipline, interrupting it to march on the enemy, resuming it, if necessary, more violent than before to clinch by bullets the demoralization caused by his advance.

The same preoccupation of giving to fire all its power is constantly evidenced by the Prussian infantry of 1870, although its armament is then inferior to the French "chassepot."

One seeks first of all to reach a position from which the enemy can be held, crushed by fire, while behind this position is being organized the assault, which by its advance uses and develops the acquired superiority.

The position is reached by small units, extended in any possible way, taking advantage of all natural cover, the primary consideration being to reach there without suffering from the enemy's fire.

Once that position is occupied, the battle for superiority of fire is begun, every effort being made to maintain both fire control and fire direction (objectives and ranges). This lasts as long as may be necessary to prepare the attack, which is normally carried out chiefly by troops advanced in echelons from the rear, reinforcing the units in the firing line.

To-day we find the same preoccupation of efficient fire control appear in the German maneuvers. Von der Goltz writes:

"Another phenomenon which does not lack importance has also been observed since the war of 1870. I refer to that new principle concerning infantry, principle in accordance with which that branch of the service endeavors to *control its fire by stricter discipline* than in the past, *in spite of the extended formations* which characterize its mode of fighting."

The battle for superiority of fire has become an inevitable part of the attack. We can no longer, as was done in the past, charge an enemy intact. If, therefore, in accordance with the argument of Souvarow: "The bullet is mad, the bayonet alone is intelligent," we admit that the assault is the supreme and necessary argument to complete the demoralization of the enemy and put him to flight, it nevertheless is true that superiority of fire is an indispensable advantage to obtain first.

The strongest of moral qualities disappear under the effects of modern weapons if the enemy be allowed to make full use of them. The attack will inevitably be halted if the question of superiority of fire has not been previously settled. Such superiority alone allows further progress, because it deprives the enemy of part of his resources, affects his morale, reduces his numbers, uses up his ammunition, flattens him on the ground and makes him incapable of any sound and complete use of his weapons.

But since the battle by fire has become a necessity, we must prepare it and study it in times of peace, determine the results to be sought and the methods to be employed, decide how control will be exercised and what can be expected of the troops.

But we must first recognize that the work of the infantry cannot be the mere processional development of the two means of fighting which it possesses: advance and fire, from 1,200 meters up to the enemy's position.

Infantry cannot expect to carry out the necessary and decisive fire battle with men shooting more or less independently as soon as they reach 1,200 meters, nor can it assault with these same men. Its formations at that range, the use it makes of its forces, must aim therefore at the preparation of the first act, the battle for superiority of fire, considering its possibilities in the same way as they consider the possibilities of the later assault with the bayonet.

Thus we find in the infantry battle, more or less altered by modern conditions:

(1) A period of marching to attain a position allowing fire of assured efficiency, or the nearest position that can be reached in safety. During this period the troops cause little harm to the enemy, and are exposed to heavy casualties unless they avoid them by:

(a) Formation; a poor help in view of present armament; the least vulnerable formations are still too much so to allow of marching;

(b) Some limited fire, able in spite of its small efficiency to maintain a certain confusion among the enemy and to a certain extent paralyze his means;

(c) Ground and the cover it affords; that is the only real protection against enemy fire, and it therefore decides what formations to employ; they must allow the use of cover, but also in view of the battle for supremacy of fire to be soon undertaken, these formations must avoid dispersing the troops, losing their cohesion and letting them waste their ammunition.

(2) A period of battle for supremacy of fire. That period requires different qualities from the men and from their chief. It requires from the former the power to keep up during ten, twenty, thirty minutes or more an efficient fire increasing in violence, yet never wild or

wasted. From the chief it requires knowledge of the results sought, of the technical means of obtaining them, of the ways to control and use fire under all conditions.

(3) A period of assault which will be discussed later.

From the necessity of the battle for supremacy of fire has resulted the need of instruction camps permitting a study of fire direction, fire control and fire discipline, a sound understanding of the use of fire in war, and the practice of fire by the troops.

How the Attack Would be Carried Out To-day

The tactics employed by the 2nd Battalion of the 37th would no longer be sufficient to-day. In the presence of a serious opponent it would be necessary, as I have already explained, to hasten to the strong points, and to occupy them if the enemy had not already done so.

If the enemy were found there, he should be immediately attacked, and these points strengthened as soon as they were taken.

But let us return to the attack. How should it proceed in order to assault the Prussian advance guard on the positions it occupied June 27th, 1866? How could it use, on the ground that leads to Wenzelsberg, the theory of the use of fire and of advance which I have expounded previously?

We presume the troops ready and assembled at Schonow. Their purpose is to thrust the enemy back from the positions he occupies, and to take his place there. To that end, the positions must be reached by following a direction and a road leading to them. That direction and road cannot be chosen without care, for if we merely march on the enemy without protection, we shall melt under his fire, and either fail in assaulting him or assault badly.

The first requirement, therefore, is to seek roads that enable us to advance under cover from enemy fire for as long a time as possible, until being, we also, able to use our weapons we can deal with the opponent on at least equal terms.

Small defiles and natural cover can nearly always be found when sought. In this case, the ground south of Wenzelsberg provides: the hollow roads leading to the Evangelical Chapel, then the orchards and village, then the big ravine southeast of the village, very easy to reach.

Under these various forms of cover we can, by various methods, push small columns, easy to handle and able to take advantage of the ground, assembling thus important forces without loss.

Such is the end of *formations* and rigidity. They disappear because by themselves they guarantee nothing. The desired result being known, to approach as near as possible to the enemy under cover from his fire, the regulations provide means, variable according to circumstances, of doing so: column of sections, of platoons, of companies, lines of sections, etc., the choice depending on individual circumstances. But as long as the troops are advancing thus, they are unable to fight, to receive the enemy effectively if he appears. They must therefore be protected.

Hence the preliminary occupation of strong points covering these places of assembly. In the present case, Wenzelsberg, the triangular wood, the Maison Forestière, Sochors.

It is a case of the advance guard in battle, insuring

and pushing as far as possible the preparation of the attack.

As a matter of fact, we can see here:

The 1st advance guard battalion occupying at once Wenzelsberg and the triangular wood;

The 2nd battalion holding the Maison Forestière, Sochors, the neighboring wood;

Further back, the brigade bringing:

Its first regiment in the ravines of Wenzelsberg and of the Maison Forestière;

Its second regiment massing north of the Meier farm, and pushing one company into the wood west of Wenzelsberg for the purpose of keeping the possibility of action in that direction, if it become necessary.

After we have thus prepared and assembled the attacking forces behind the natural cover available, nothing is accomplished unless we actually assault the enemy.

For that purpose we must reach him, but the ground no longer offers any possibility of advancing without risking his blows. It is on our arms that we shall depend in order to smooth the difficulties of the road, employing them under such conditions of numbers, of time and of space as enable us to produce more effect (material and moral) on the enemy than he can produce on us. From now on, it is under cover from fire that we propose to advance. The men will extend, occupy with many rifles every point from which the objective can be swept; the extremity of the chosen means of approach becomes the principal fire position, to which we shall give all possible development in order to create an advantage in our favor.

We have already seen how the battle was organized for superiority of fire. Under fire the man obeys such

THE ADVANCE GUARD AT NACHOD

leaders, officers or section commanders, as he knows, and the leaders in turn should know every individual with whom they have to deal.

First of all, the objective of the attack must be determined. The same considerations as to space to be covered under enemy fire and as to the effects sought result in the choice, as first objective, of that enemy point which is nearest at hand, and on which can be employed a numerical superiority guaranteeing a superiority of results. In this case, we start from the triangular wood and the northern part of the ravine to attack the wood east of Wenzelsberg: objective close at hand, weakly held, forming a salient, easy to envelop and to assault with superior forces.

The 3rd Battalion extends (covered by the 1st Battalion of the advance guard, which abandons to the 2nd Regiment the total or partial occupation of Wenzelsberg) in the northern part of the ravine, and begins at these points the battle for superiority of fire.

It prepares, at the same time, to carry the wood, and for that purpose to march without losing its superiority of fire. It will proceed in extended formation supported by columns of companies or by troops in line.

Once the wood is reached, as the decision inside may be delayed, it is sought simultaneously outside by troops operating outside its fringe.

To enter the wood and conquer it (because of the length of its fringe and of the distance to cross before reaching it) is a possible task for the 3rd Battalion; it will debouch from the northern part of the ravine, organized:

(a) To first strike the salient aimed at, which will

probably require 2 companies in extended formation, a third one supporting in the rear;

(b) To maneuver along the fringe, overcome if necessary all resistance inside the wood, which will require the fourth company in reserve, forming echelons to the right.

This attack is supported as to its front by the 1st Battalion, employing part of its forces along the northern fringe of the wood, insuring at least the occupation of that wood.

Yet the attack may be powerless, stop when insufficiently pushed: the 4th Battalion moves forward in regimental reserve inside the ravine, ready to interfere, either to support the exhausted attack or to follow it up once it has succeeded.

But in order also that the attack may succeed, it must be protected against any surprise from the enemy, counter-attacks or powerful fire, especially on its outer flank. That duty belongs to the 2nd Battalion (previously of the advance guard) which, in order to carry it out, occupies the northern fringe of the wood northeast of the Maison Forestière with 1 company, afterwards occupying with another company the northern end of the ravine and the little wood touching it; it thus covers the flank of the attack, advancing constantly without abandoning the strong points in the rear.

That effort requires that the 1st Regiment move north of the Maison Forestière, ready to act with unity, although it engages only one battalion on the objective sought.

The wood east of Wenzelsberg being thus carried, its possession is at once assured by occupying its northern fringe with troops, behind which are reorganized the more or less broken up troops which have assaulted; a new objective is afterwards chosen and attacked in the same manner, from the most suitable direction, with a special distribution of the troops adapted to the new conditions. In such a series of successive actions is a battle transformed with modern weapons: an effort to advance on the right when no further frontal advance is possible; to maneuver by one flank when the other is stopped; keeping always the freedom to maneuver by préliminary measures of safety; preserving always, at the point of attack, the possibility, allowed by the nature of the ground, of applying numerical superiority obtained by careful planning.

In order to be sure of getting the wood, our point of attack, we shall use there all the forces we have, all the rifles and all the guns. We must ask the artillery to prepare the attack, and for that purpose to place itself where it can observe (or obtain indirect observation). But the hostile artillery will observe it also and interfere with it, so that it should take a position from which it can observe without being observed, or else obtain a superiority over that enemy artillery. For that purpose it also requires numbers.

The Wenzelsberg wood being seized, the plateau cannot any longer be held by the Prussian advance guard. The attack of the second position could then be undertaken. From what precedes, we find in an operation the following characteristics:

The idea of action, coming from the mission assigned to us or from the situation known to us; in this case an idea of attack;

The idea of attack being adopted, its direction must be sought, it must be prepared, carried out, protected.

The direction depends, as we have seen, on the most favorable ground, the one that offers best cover and from which we can use our rifles and guns together and against the same objective.

As to the three conditions of preparation, execution and protection, we have seen what they implied. In any case, there is but one offensive to be undertaken at a given time, with but one objective.

All the forces employed for that operation, infantry, artillery, cavalry, engineers, are detached only to the extent absolutely necessary for protecting, preparing and assuring the operation, that is for insuring constantly to the main body:

Freedom of action;

Liberty to dispose of its troops;

Strict economy of forces.

The main body, moreover, seeks constantly to employ in the attack forces insuring to it an undoubted superiority, to use a minimum of troops for the mere purpose of protection.

The Infantry Battle

About 10.30, Steinmetz orders General Loewenfeld to remain on his position. The main body, the artillery reserve in particular, is ordered to hasten its advance. However, the Prussian advance guard is worn out and in a dangerous situation.

On the Austrian side, the Jonack Brigade has arrived; it has massed south of Domkow, and will advance to attack the position's right flank.

About 9.30 this brigade moves forward:
Wasa Regiment in the front line;
Prince Royal of Prussia Regiment in the second line;
Clam-Gallas Regiment of Uhlans in the third line.

The whole is protected by its battery, east of Domkow towards the northern access to Schonow. The two Prussian batteries open fire on it at 2,300 meters.

The brigade moves forward behind its band, and arrives north of Schonow about 10 o'clock or 10.15. The 14th Battalion Chasseurs comes on its right, having previously protected the advance against any enemy enterprise from the direction of Giesshübel.

At the same time, the Rosenzweig Brigade masses opposite Prowodow, later moving east of that locality, with its advance guard (17th Battalion Chasseurs) on the heights between the Kleny road and Wenzelsberg.

About the same hour the Waldstätten Brigade received the order to move from Skalitz on Wysokow.

At 10.30 the Jonack Brigade begins to attack, while the Hertweck Brigade retires on its right, and five squadrons of cuirassiers on its left hold the Skalitz road, below the heights. The brigade moves again in the following formation:

The Wasa Regiment is in the front line, its three battalions on the same level, the 14th Battalion to the right of that line; the Prince Royal of Prussia Regiment forms the second line. The brigade advances towards Wenzelsberg; as it follows the Hertweck Brigade, stragglers from that brigade soon force it to slow up, and even draw with them part of its front line. It is protected neither to the right nor to the left, so that it is soon attacked by detachments of Prussian infantry which, debouching from the neighborhood of the Maison Forestière and of Sochors, fire on the brigade's flank.

The 14th Battalion Chasseurs, then the 3rd Wasa, the 1st Wasa and the 3rd Prince Royal of Prussia face right to meet this attack, and stand at the foot of the ravines which surround Sochors. The attack is continued, but with three battalions only.

It is with such reduced forces, without serious encounter with the enemy, that the Wenzelsberg Chapel is reached. From that point action is undertaken on the Wäldchen, in which participate the 25th Battalion and the 2nd of the Kellner Regiment, which previously occupied the northern fringe of Wenzelsberg, together with the 17th Chasseurs, advance guard of the Rosenzweig Brigade.

Under this enveloping attack from six Austrian battalions, the Prussians lose the southern part of Wäldchen.

Meanwhile, the Rosenzweig Brigade has started, with the Gondrecourt Regiment in the front line and the Deutschmeister Regiment in the second line, marching towards the western extremity of the wood.

General Jonack calls on it for reinforcements, and the 1st and 2nd Gondrecourt Battalions are instructed to support his left. There are consequently eight Austrian battalions coming to strike that weak point of the Wäldchen, carrying it and plunging into a wood 300 to 400 meters wide with an average length of 1,300 to 1,400 meters. Naturally they are lost in it without serious results.

The remainder of the Rosenzweig Brigade (2nd Gondrecourt Battalion and Deutschmeister Regiment) reached, shortly after, the southwestern point, where also arrived the brigade's battery. These battalions, marching towards the Wenzelsberg Church, strike the 2nd Battalion of the 37th. The latter meets them with very energetic fire which stops the scouts and the 2nd and 3rd Deutschmeister Battalions. But it is turned itself by the 1st Deutschmeister and by detachments which debouch from the Wäldchen.

It soon retires on Branka-Wald, turning about several times to carry out rapid fire; it is met by one half-battalion (Bojan of the 3rd of the 37th) which tries a counter-offensive, and by another half-battalion (Suchodoletz of the 58th) which has hastened from Alstadt and established itself on the western fringe of Branka-Wald.

The Prussian cavalry and artillery can hold out no longer. They withdraw behind Branka-Schlucht.

The movement of retreat in the center enforces a similar movement on the part of the troops at Maison Forestière and Sochors, which fall back slowly towards Branka-Wald. At Wysokow there has been no attack.

The whole Prussian advance guard is engaged, and it is beaten. But it is almost *noon*.

Only a little while more need the resistance be prolonged. The infantry is disorganized, the cavalry must be called upon to continue the battle.

The enemy takes advantage of his success to occupy the approaches of the Neustadt road; he must be stopped at any cost.

CAVALRY BATTLE (11.30 TO NOON)

The ground between Wysokow and Wäldchen offers a road, lined with trees, which runs from Wysokow to Branka-Wald and outlines the summit of the slopes which rise from the meadows. These slopes are crossed by hollow paths and ditches, and are covered by the harvest and by a few small woods.

Meanwhile, the brigade of Wnück, hastily called, has crossed Nachod at 10.30. It reaches the Neustadt road, and takes up a position of assembly behind the crest. It is formed in line of columns of squadrons, the 1st Uhlans on the right, 8th Dragoons on the left. A troop of Dragoons patrols in front. A squadron of Dragoons (2nd of the 4th) scouts in front of Wysokow; it soon falls back before the superior forces of the Austrian cavalry. It is the moment when the Prussians are thrown out of the Wäldchen. The battery of the Wnück Brigade establishes itself towards the Neustadt road.

Opposite, the Solms Cavalry Brigade holds the front of the meadows west of Wysokow; it is near the bridge on the road.

That brigade's battery seeks first of all to take up positions between Wysokow and Kleny, and to sweep the ground with its shells. It afterwards advances as far as the hay-lofts of Wysokow without being able to occupy positions. It produces no effect.

General Raming orders General Solms, at the time the 17th Battalion Chasseurs enters the Wäldchen, to advance on the plateau and protect the left flank of the Austrian infantry. It was about 11.30 when the squadron of Prussian Dragoons (2nd of the 4th Dragoons) retired before the Austrian cavalry and announced the advance to the Prussian infantry in Wysokow (Kurowski half-battalion of the 3rd of the 37th) and to General Wnück. At the same time, and by his patrols, General

THE ADVANCE GUARD AT NACHOD

Solms learnt of the presence of a Prussian Uhlan regiment on the plateau.

The Austrian cavalry, its left resting on the railroad, attacks.

First the 5th squadron of the Archduke-Ferdinand regiment skirts Wysokow, debouches from the hay-lofts, wheels right and charges; General Wnück immediately moves forward, and deploys at a trot, the regiment of Uhlans.

The 1st Squadron of Uhlans makes a frontal attack, the 2nd Squadron attacks on the flank, followed by 3 troops of the 3rd Squadron.

The encounter takes place at the southern edge of Wysokow, near the eastern outlet.

The 2nd Squadron of Hessian Cuirassiers perceives what is going on, and throws itself into the struggle when the 2nd Squadron of the 4th Prussian Dragoons arrives and attacks in its turn, debouching from the eastern outlet of Wysokow. There are altogether 2 squadrons of Austrian Cuirassiers against 3¾ Prussian squadrons.

The struggle becomes very confused. In spite of their numerical inferiority, the Austrian Cuirassiers vigorously keep up the engagement; being thrust back, they reform themselves and again start to attack.

But the Kurowski Half-battalion (of the 3rd of the 37th) has been warned of the approaching enemy cavalry, and it has advanced towards the eastern part of Wysokow, and is ascending rapidly the slope to the south of the village. The half-company Chasseurs, at that moment in the middle of the village, immediately moves to the southern edge.

The Kurowski Half-battalion has not finished its movement when the charge begins. It hastens to the crest, for it cannot see clearly, and it forms there a hollow square, its front and left flank firing on the mass of Austrian Cuirassiers who seek to rally at the foot of the slope. The latter are thrown into confusion, and soon pushed back. They again try to rally under cover of the haylofts, but come under fire of the half-company of Prussian Chasseurs, and finally retire.

The decision in the cavalry battle was undoubtedly due to the intervention of the Prussian infantry, which had known how to seize a good opportunity to intervene. It was a triumph of solidarity, of initiative on the part of the commanders of small units, brilliant evidence of the warlike activity found in the ranks of that army which instruction animates and inspires.

General Wnück, ignorant of the Prussian infantry's presence at Wysokow, believes the village to be occupied by enemy infantry. He orders the assembly. He begins no pursuit.

But a battle was going on at the same time near the Wäldchen.

The two-and-a-half Hessian Squadrons placed as we have seen on the Austrian line's right have followed in echelons advancing along the ground's depression west of Wäldchen towards the northeast. The commander of the 4th Squadron of Prussian Uhlans has barely noticed this move before the Austrian squadrons deploy suddenly at a gallop and charge him. The captain meets the attack with his squadron and two troops of the 3rd Squadron on the left.

The encounter takes place near the northeastern end of the Wäldchen.

The Prussian Uhlans are enveloped by the Austrian Cuirassiers; being at close quarters they cannot use their lances. Part of them are hastily withdrawn to the Neustadt road.

The position of the Prussian Uhlans was already fairly critical when intervention occurs from the 3rd Squadron Hessian Cuirassiers which being attached to the Rosenzweig Brigade had placed itself behind the Wäldchen. The squadron skirts the edge of the wood and throws itself at the flank of the Prussian Uhlans.

General Wnück joins the 8th Dragoons which then moves toward the northeastern point of Wäldchen and attacks the Hessian Cuirassiers without letting himself be stopped by fire coming from the woods.

The 1st and 2nd Squadrons make a frontal attack, the 3rd and 4th envelop the Austrian right. A new struggle takes place, but the enveloping tactics and superior numbers of the Prussians (five-and-a-half squadrons against three-and-a-half) give them the decision. The Austrian Cuirassiers retire slowly and the Prussian Uhlans return behind the crest.

The Austrian infantry in the midst of the tumult that reigns in Wäldchen does not suspect the cavalry battle going on close at hand and it does not intervene; it is incapable of the happy initiative shown by the Prussian infantry of Wysokow.

General Wnück has no reserve left at all. He knows Wäldchen to be occupied by the enemy. He believes Wysokow to also be in his possession. He fears to meet numerous and fresh enemy forces. At the moment when he appeared on the battlefield, the fight was going on along the whole front; the main body of his army corps was approaching. He cautiously decides to reassemble his brigade and stop all pursuit.

Such are the reasons given by him, while claiming a victory, to justify his timid behavior.

The Solms Brigade was assembling west of Wysokow; later it withdrew on Kleny; it will not appear again that day.

The Prussian brigade resumes its original position. Later on, the Austrian artillery's fire will compel it to withdraw behind Branka.

ATTACK BY THE AUSTRIAN INFANTRY OF THE BRANKA WOOD—NEUSTADT ROAD POSITION (NOON)

About the same time when occurred the cavalry engagements which we have just considered, the Austrian infantry sought to debouch from the Wäldchen to attack the Branka wood and Neustadt road.

At first, enemy fire easily masters these disconnected efforts. Moreover, the Prussian Half-battalions of Bojan (of the 3rd of the 37th) and of Suchodoletz (of the F of the 58th), held until now in reserve, come out of the Branka wood, enveloping with a powerful fire the two Austrian battalions (1st and 2nd of the Prince Royal of Prussia) which have succeeded in debouching from the northeastern point of Wäldchen, and also the 17th Chasseurs. This fire being insufficient, the two Prussian battalions leap 200 paces forward and, at a distance of 350 paces, fire several volleys. At the same time, General Wnück has ordered the 8th Dragoons to stop the in-

THE ADVANCE GUARD AT NACHOD

fantry which is debouching from Wäldchen. The result is a series of cavalry attacks against the Austrian Chasseurs and against the battalions of the Prince Royal of Prussia Regiment.

The companies of the 17th Battalion of Austrian Chasseurs that are not engaged immediately retire in the Wäldchen. The others form a square to resist the cavalry charges, but they suffer considerable casualties under the fire of the Prussian Half-battalions (Bojan and Suchodoletz). Compelled to deploy, they are charged and soon sabered back into the wood.

The two battalions (1st and 2nd) of the Prince Royal of Prussia Regiment, and the detachments of the 25th Battalion and of the 2nd Kellner, endeavor to support them but suffer the same fate. A flag is lost.

The Bojan Half-battalion takes advantage of this success to rush the Wäldchen, supported shortly by the first troops of the army corps' main body, arrived at last and having taken part already in the battle for supremacy of fire against the Austrian 17th Chasseurs Battalion.

During the fight, the Rosenzweig Brigade has engaged its reserves (1st and 2nd Battalions of the Gondrecourt Regiment) through the region south of Wäldchen against the position on the Neustadt road. The repeated attempts of the Prussian cavalry, generally restricted to threats on that road, compel these battalions to also form into squares. As a matter of fact, they suffer little from the cavalry, but they melt under direct fire from the fringe of the Branka wood and under flanking fire from every direction. The same is true of the Rosenzweig Brigade which has sought to occupy positions before Branka Wald.

At 12.30, the space between the Neustadt road, Wäldchen and Wenzelsberg was abandoned by the Austrian infantry, and the attack of the left flank, furnished by the Jonack and Rosenzweig Brigades, and supported by part of the Hertweck Brigade, had failed.

All progress by the Austrians had been stopped in front of the second position through the combined work of the enemy infantry and cavalry. No attempt had been made against either flank of the Prussian advance guard.

The critical period was over for its army corps. The battalions of the main body were beginning to deploy on the plateau. The Austrians had no fresh troops left, except the Waldstätten Brigade, the reserve artillery and the Schindlöcker Cavalry Brigade. But they were ready for action. An energetic counter-offensive would still have permitted the head of the 5th Prussian Corps to be thrust back on the defile. It seemed unlikely, however, that where three Austrian brigades had failed against a weak advance guard, one brigade and some demoralized troops could reverse the situation.

REMARKS ON THIS PHASE OF THE BATTLE FROM 10.30 TO NOON

If we consider this part of the battle, directed by General Raming, we find that it calls for some remarks.

To begin with, at the time that he takes over command of the battle, the Hertweck Brigade is already defeated, the Jonack Brigade is on the point of engaging itself. The Rosenzweig Brigade will soon reach the scene, while the Waldstätten Brigade and the artillery reserve are still in the rear, at Skalitz.

THE ADVANCE GUARD AT NACHOD

This dispersion of his means of action is evidently unfavorable to battle. The most unfortunate thing possible would be to make it worse. All forces are already unable to strike simultaneously, the worst thing would be to thrust them drop by drop into the action. Moreover, General Raming cannot ignore what has just occurred.

The Hertweck Brigade has suffered heavy casualties, and is in full retreat;

The enemy has shown himself on a front of 3,000 meters;

The Austrian artillery has displayed its inferiority;

The Prussian cavalry has shown several squadrons.

How then can a new action be contemplated with one brigade against an enemy whose morale has improved? Only with all his forces together, working in unison, could Raming have again undertaken action.

To utilize the remnants of the Hertweck Brigade, to push forward together the Jonack and Rosenzweig Brigades, to form a reserve with the Waldstätten Brigade hastily summoned by the Skalitz road: these were in a general way the measures to be taken.

But how can the troops now be thrust into the attack? We have seen the results of a battle conducted in a purely frontal manner, with no idea of combination. It is evidently a flanking maneuver that should be combined with a frontal attack.

On which flank should he act? On the most favorable one, the one which first enables the attack to develop to the best advantage, and which afterwards promises the most decisive results.

These two conditions were found in the Wysokow-

Altstadt direction. An examination of the map, the sight of the ground, made that clear.

In the region north of Wäldchen was ground practicable for all arms. No other obstacle than Wysokow. This locality and the region further north could be reached under cover.

As a final result: the attack north of Wysokow, on the Wysokow-Altstadt space, took the enemy in the rear, defeated all his attempts against the plateau of Wenzelsberg, and closed finally the opening of Nachod.

In accordance with these ideas, the plan should have been to:

Intrust to the Hertweck Brigade the occupation of Wenzelsberg and of the regions further south;

Thrust the Jonack and Rosenzweig Brigades into the interval Wäldchen-Wysokow and north of Wysokow;

Prepare that attack with all available artillery;

Protect it on the north with a large body of cavalry;

Bring the Waldstätten Brigade in reserve to Kleny.

In any case, the attack should avoid the region of Wenzelsberg, a wooded, uneven region which compelled it to break up into small detachments, to waste itself in powerless efforts, impossible to coördinate, in which no advantage could be taken of numerical superiority.

To justify his attack by the plateau of Wenzelsberg, General Raming has claimed that he did not know whether the Prussian attack was taking for its objective the road of Neustadt, that of Skalitz or that of Wenzelsberg. But no maneuver should consist in merely responding to the attempts of the enemy.

Every maneuver must be the development of an idea; it must have some purpose: in this case, to close the opening of Nachod, and with that object to take the direction that leads there most easily, most safely, deciding then all the dispositions of attack.

At the same time, the Jonack Brigade becomes engaged without reconnoitering the position of the enemy and his situation. It offers a thin extended line followed by battalions in close order, unable to maneuver; it advances over the ground by which flee some of the men of the Hertweck Brigade. It is not protected on its right flank. It has left its cavalry behind, and its artillery supports it only from afar. The results will soon be seen. The extremities of the line are more or less dragged into the flight. To prevent a counter-attack, four battalions or so go off at an angle. Numerical advantage has disappeared. Not a shot has been fired, yet out of 7 battalions the attack has only 3 left.

We must also note the coöperation of the Kellner Battalion and 25th Chasseurs in the attack of Wäldchen, which by abandoning Wenzelsberg, the only strong point available, exposed it to be seized by the enemy if his temporary numerical inferiority had not prevented it.

Again, in the attack on the second position, higher direction is lacking, there is no unity. How could it be otherwise? Eight battalions trying to debouch on the eastern edge of Wäldchen, measuring less than 400 meters. Further south, chance has led before the Neustadt road, in compact masses, the reserve. It has no plan of battle, but acts blindly and no communications are even kept up.

Better led, the Austrian attack would have succeeded.

But what would have been its results on later operations?

The 5th Prussian Corps, finally thrown back, would have been unable to debouch from Nachod. If we note that, on the same day, the 1st Corps met with a serious reverse at Trautenau, we find that entry into Bohemia would have been impossible for the 2nd Army. What became then of the plan of General Moltke? Once more we find strategy, however brilliant, at the mercy of tactics.

As regards the cavalry battle south of Wysokow, both sides have claimed a victory. Both may be right if we consider only that battle in itself. In reality, the Austrian cavalry showed itself very brilliant, well drilled and of undoubted merit. It was not properly commanded. The Prussian cavalry was more cautious, less thoroughly trained: it had no such speed or piercing power. But it was efficiently handled. It displayed tactical ability. As far as results are concerned, it is victorious. Two cavalries do not fight just for the purpose of finding out which is the best. There is always a general situation to be considered, a tactical purpose to be carried out. For the Austrian cavalry, as for the Austrian infantry, it is in this case to reach the opening of Nachod; it did not do so. For the Prussian cavalry, as for its infantry, the aim was to cover the opening; it succeeded in doing so.

Among other mistakes, the Austrian cavalry does not protect itself in front or towards Wysokow, and a decisive surprise is the result. The employment of cavalry is also different: General Steinmetz has on the field all his cavalry (about 12 squadrons); the Austrian commander who has over 30 squadrons available does not engage more than 5. He has a superiority of numbers, yet it is by numbers that his cavalry is defeated.

THE ADVANCE GUARD AT NACHOD

ARRIVAL OF THE MAIN BODY OF THE 5TH PRUSSIAN CORPS, AND OF THE AUSTRIAN WALDSTATTEN BRIGADE

The 5th Army Corps has left at 5 o'clock its Reinerz quarters; it has halted near Gellenau at 8 o'clock, has resumed the advance at 9 o'clock, receiving at that moment a new invitation to hasten towards Nachod.

The order of march presented no peculiarities, unless it be that the artillery was distributed by batteries along the whole length of the column.

The bridges had been reconstructed on the Mettau, and two more built; all were to be crossed with caution, which much delayed the column's progress.

General Kirchbach, commanding the 10th Division, has preceded the column on to the field of action; he has watched the loss of Wäldchen and the cavalry engagement.

As soon as his troops arrive, he orders the General in command of the 19th Brigade to recapture and occupy Wäldchen, while he proceeds himself towards Wysokow, which the commander of the army corps has ordered occupied. A half-battalion of the F of the 46th therefore advances towards Wäldchen at the moment when the 8th Dragoons charges and when the Bojan Half-battalion resumes the attack. It is followed by the 2nd of the 46th which, with the Bojan Half-battalion, enters the Wäldchen, whence they both soon eject the Austrians.

Under the protection of these troops advances, behind Wäldchen, what is left of the 46th, while the Division Commander sends the remainder of the 19th Brigade, nearly the whole of the 6th Regiment, towards Wysokow.

His original idea was to move there the 20th Brigade. But he fears to have no time to do so; in haste he sends the 6th Regiment. In the Wäldchen the Prussians reach the southern fringe and carry the church of Wenzelsberg. The retreat soon spreads to the center and right of the Austrians, though they have not been attacked by the Prussians.

After an engagement of half-an-hour, the Rosenzweig and Jonack Brigades abandoned the ground from Wäldchen to Sochors, which had been so difficult to conquer.

While occurred, about noon, this fight on the western slope of the plateau, and while the crisis was ending in favor of the Prussians, there arrived on the Austrian side the Waldstätten Brigade; it placed itself astride of the Skalitz road.

At the same time entered into action, east of Kleny, the Austrian artillery reserve, with 2 batteries of 8 at the south, 3 batteries of 4 at the north of the road; it opened a very efficacious fire against the plateau, and rendered very difficult the installation of Prussian batteries arriving one behind the other, and also the advance of the enemy infantry. The Wnück Cavalry Brigade was compelled by that fire to retire to 500 meters east of Wysokow, near the road.

On the other hand, thanks to that fire, the retreating Austrian infantry could reorganize easily near Prowodow and Schonow.

The battery of the Rosenzweig Brigade after abandoning its position at the Wenzelsberg church, was appearing again on the heights north of Domkow.

When General Raming sees, from his position in advance of Kleny, on the main road, the movement in

THE ADVANCE GUARD AT NACHOD

retreat of the Jonack and Rosenzweig Brigades, he orders the battalions of the Waldstätten Brigade nearest to the road (2nd of the Hartmann Regiment and 3rd of the Franck Regiment) to attack the fringe of the wood.

The 2nd Hartmann leaves, and is greeted by the Prussian Half-battalion holding that part of the fringe with a violent fire which throws its leading men, advancing in extended order, back on those that follow. The Prussians, moreover, advance into the ravine which leads to Provodow, and open again a strong fire on the 3rd Franck Battalion which was continuing the advance; the result is the same. All the Prussian companies near the scene of this action take part in it successively, and the whole Prussian line soon even reaches the hollow of Schonow and Provodow where it stops, apparently by order of the Brigade Commander.

At the same time as the 19th Infantry Brigade was arriving, the Prussian main body's artillery sought to take up positions between Wysokow and the Wäldchen, partly protected by the Wnück Brigade. The attempt failed under the fire of the powerful Austrian artillery.

About noon one battery appeared, then retired; a quarter of an hour later a second one met the same fate; three quarters of an hour later, a third one.

Two other batteries, arrived later, succeeded in establishing themselves firmly, but near the church of Wenzelsberg.

Here ceases the success of the Prussian artillery, except for the fact that the first three batteries ended later by establishing themselves between Wysokow and Wäldchen, after the Austrian artillery had been moved elsewhere.

ENVELOPMENT OF THE PRUSSIAN RIGHT BY THE WALDSTATTEN BRIGADE. ENTRY INTO ACTION OF THE 20TH PRUSSIAN BRIGADE

At one o'clock, General Raming orders General Waldstätten to attack Wysokow. The latter had advanced with the main body of his brigade (before the attack of Wäldchen by two of his battalions) on the road, and had afterwards turned to the left so that at 1.30 he reached the hollow between Starkoc and Wysokow.

On the order of General Raming, three batteries of the artillery reserve proceed towards the height north of Wysokow, in support of the Waldstätten Brigade, while the 11th Cuirassiers advances towards the same height with the object of enveloping the Prussian right.

The village of Wysokow is built over a stretch of 2,000 meters, along the road which runs from Branka towards Skalitz. It extends to a fairly steep ravine, the southern slope dominating the northern. Both ends of the village contain more houses than the middle.

The ground south of Wysokow is quite different from the ground on the north. Here the plateau of Wysokow dominates the whole country, commands the flank of the Wenzelsberg Plateau, and holds under its fire Nachod and Altstadt. It is easy to defend facing east.

The importance of the plateau of Wysokow on the day of the battle has escaped the Austrian General Staff. That seems strange, for from the position east of Kleny, where this General Staff was, the panorama is striking.

Did they wish to stop up the opening of Nachod? It was the heights south of the road, or those to the north,

THE ADVANCE GUARD AT NACHOD

that should have been seized. The heights to the north were preferable, because from them one could more quickly reach the Nachod road, and one dominated Wenzelsberg.

In any case, General Raming has ordered the attack on Wysokow and an envelopment of the Prussian right. General Waldstätten disposes for that attack of four infantry battalions (6th Chasseurs, 1st Hartmann, 1st and 2nd Franck), the brigade battery, 3 batteries of the artillery reserve and 1 regiment of Cuirassiers.

The brigade battery arrives first on the height of Wysokow, protected by a detachment of Chasseurs; then a battery and a half of the artillery reserve which takes up positions on the plateau of Wysokow; another battery is near the road, and the remaining half-battery on the low height of the railroad; the regiment of Cuirassiers is on the right of the battery and a half.

From the Prussian side, General Kirchbach has observed the movements of the Austrians to the north. He sees the danger which threatens Wysokow, and realizes that the occupation of the locality by two battalions will not suffice to hold it. He calls to that point the 20th Brigade.

This brigade has had to slow up considerably for the crossing of the Mettau bridges; it has arrived too late on the plateau; it has been sent behind the Wäldchen as reserve. Besides, two of its battalions have been kept in reserve at Altstadt on the firm order of the Prince Royal, and a third battalion accompanies the batteries stuck in the Mettau's mud, so that the brigade consists temporarily of only three battalions (1st and 2nd of the 52nd, F of the 47th). In such condition it is recalled towards

Wysokow from its march on Wäldchen; it will reach Wysokow too late for the opening of the attack.

The latter begins about 2 o'clock; it is made against the Prussian troops distributed as follows:

At the western end of Wysokow, one-and-a-half company Chasseurs and two-and-a-half companies of the 6th; altogether the equivalent of one battalion;

In the middle of the village, about one battalion holds the road and height of Wysokow;

At the eastern end is one-half battalion.

One battery appears at the north of the village, but cannot hold there more than half-an-hour.

The Austrian attack develops against the western extremity and the center of the village.

Part of the 6th Chasseurs, taking advantage of the deep cut which runs north of the Wysokow height, moves against the northwestern fringe, a few detachments enter the village: result of advancing under cover.

Another portion of the 6th Austrian Chasseurs extend and attack further north, enveloping the Prussian right. At that moment, General Waldstätten advances his three other battalions (1st Hartmann, 1st and 2nd Franck), partly against that portion of the village which is already occupied, partly further east.

The Prussian situation is most critical. However, the commander of the 20th Brigade, General Wittich, arrives and immediately calls the three battalions to which I have referred (1st and 2nd of 52nd, F of 47th).

The Division Commander orders him to cross Wysokow, to reach the heights on the north, to attack the enemy there and thrust him back. He is informed at the same time that the Wnück Cavalry Brigade will protect

THE ADVANCE GUARD AT NACHOD

his right flank. One battalion, however (1st of 52nd), is employed to reinforce the occupation of Wysokow.

There remain, for counter-attacking, two battalions protected by the Wnück Brigade. The ground of the plateau is little suited to infantry movements, and still less to cavalry.

The Divisional Adjutant, sent to reconnoiter the ground and guide the cavalry, has advised General Wnück that it seems easy to surprise the enemy battery located on the height north of Wysokow. To carry out this idea, the 3rd and 4th Squadrons of the 1st Uhlans are detached, and seek to reach the enemy's left flank by filing behind a fold in the ground. Later, noticing that the regiment of Austrian Cuirassiers is retiring and leaving only two troops in support of the battery, and that the Austrian artillerymen seem ready to withdraw, they start to attack, charging the two troops left by the Cuirassiers. The Austrian Cuirassiers are put to flight, and the Uhlans seize 3 guns and 6 gun-carriages.

While this maneuver is being prepared, occurs the counter-attack from the two battalions, on two lines: F of the 47th in the front line, 2nd of the 52nd in the second line.

They march to the west, utilizing the northern fringe of the village, and strike at the left flank of the attacking Austrian columns. But, as the counter-attack is immediately threatened and taken in flank by the Austrian cavalry and artillery from the plateau of Wysokow, the two half-battalions on the right, while continuing to advance, face right with their flanking companies. The Austrian cavalry does not attack. The companies then direct their fire against the Austrian artillery (one-and-a-half battery), which at the same time is shelled by the Prussian artillery. At that moment, moreover, the Austrian regiment of Cuirassiers retires by order of the Division Commander, leaving only two troops in support of the artillery. The batteries do not then believe they can hold; they are bringing up the gun-carriages, when the two Prussian squadrons appear with the success we know. Meanwhile the two infantry battalions, continuing their counter-attack, have by their fire thrown confusion in the Austrian brigade battery on the height of Wysokow; they shoot down 28 horses and 14 men, so that only 3 guns and 5 gun-carriages can be withdrawn.

Although the Prussian counter-attack has been compelled to weaken itself by these flank actions, its effect on the Austrian attack is decisive. The latter is thrown back in the greatest disorder. Those of the Austrian troops that have entered the village are still holding out in the western extremity, but the counter-attack, continuing the envelopment, succeeds in taking them in the rear, while in the front they meet the defenders of Wysokow. Disorder and retreat follow everywhere; the Austrians have definitely failed in their last effort.

General Wittich assembles his battalions and piles arms on the heights, while by order of General Kirchbach the Wnück Brigade, supported by the recently arrived F Battalion of the 52nd, takes up the pursuit.

Here again, we find the assaulting troops immediately reorganized; it is a fresh unit which is sent with the cavalry to a new enterprise, the pursuit.

The King's Grenadiers, last regiment of the army corps, were arriving at that moment.

As for the Prussian artillery, it has finally placed in position the greater number of its guns, and reduced somewhat the effect of the enemy artillery. After many unsuccessful efforts, it has only been after the arrival of the reserve artillery that all the batteries, acting simultaneously, have succeeded in establishing themselves, and even then with difficulty. To give one example only, the battery which leads the reserve artillery has lost 16 men and 18 horses before uncoupling the gun-carriages.

The 5th Prussian Corps bivouacs on the ground; in the evening its outposts stretch from Kramolna to Wysokow, and thence to the Neustadt road, with a sentry-group at Starkoc.

The 6th Austrian Corps has lost 225 officers and 7,275 men, of whom 2,500 are prisoners.

A few more remarks can be made about this end of the battle:

(1) The Prussian cavalry acts till the end. After having stopped the enemy's efforts to debouch from the woods, it attacks the enemy artillery and captures three guns; later, it carries on the pursuit. With a professional value smaller than that of the Austrian cavalry, it nevertheless knows how to do its share in the battle, how to observe the tactical principles of an advance guard; above all, it is utilized by a chief who makes it until the last produce all the results of which it is capable.

(2) The Austrian artillery has also shown itself very superior to the Prussian artillery, by its armament and the accuracy of its fire. It inflicts on the Prussian batteries, arriving one by one, losses which forbid them to carry on the conflict. However, the Prussians have won at the end of the day. The artillery battle is not, there-

fore, any more than the cavalry battle, a decisive action which finally determines the result of the engagement.

In future we shall frequently see the artillery battle remain undecided, because of distance and difficulty of observation.

(3) The engagements around Wysokow show the kind of ground to be sought for an attack.

The Austrian attack has entered the village because it was strongly supported by the artillery, that is undeniable, but also and still more because it had good means of approach which brought it under cover from enemy fire to within 300 or 400 meters from the place. A good direction of attack is one, therefore, which provides covered means of approach for the infantry and allows the use of both infantry and artillery against the same objective, with all the advantage to be obtained from numerical superiority. The attack finally failed through lack of *protection*. It will be ever thus, and any troops starting to attack must protect themselves in the directions from which the enemy may appear. The forces intrusted with that duty must, on the flank of the attack, occupy the points from which a surprise might be attempted, and also discover and repulse the counter-attack which is sure to occur.

(4) The defense of a locality consists in parrying the attack with the help of the resistance offered by the place occupied and obstacles erected there, and the counter-attack in order to crush the enemy. To neglect the counter-attack is to return to passive defense, which prevents decision and always loses in the end.

In order that the counter-attack may succeed, it must be protected, like the original attack, or it also may be

THE ADVANCE GUARD AT NACHOD

the victim of surprise. An instance is given by the two half-battalions of the 47th and 52nd obliged to face both the artillery and cavalry of the Austrians.

(5) The distribution of troops employed in the defense of a locality comprises, in accordance with what I have said: a garrison, troops of occupation in as small numbers as possible; a reserve destined to counter-attack, as strong as possible and itself supplying, at the time of action, protective troops to guard it against surprises.

The troops of occupation may be estimated on the following basis: at the moment when the enemy reaches the fringe of the village, to present one rifle per meter is sufficient resistance. Generally he can only reach one side or one part of that fringe. It is from an estimate of that portion of the fringe and the organization of a central "keep" that one must decide the number of men to be employed on the direct defense of the locality. In Wysokow, three battalions were employed.

These estimates, however, must never entail the employment in occupying the strong point of all the troops on hand. Some portion must always be kept for the counter-attack.

(6) With modern weapons showing all their power on the ground of Nachod, the Austrians suffer their greatest losses when they retire after an unsuccessful attack, or when they abandon a position which they have lost. It costs them less to advance in the attack, or to hold their ground in the defense. Hence two leading principles of modern tactics: *an attack undertaken must be carried through,* and *the defense must be kept up with the utmost energy;* there lies true economy. These principles must inspire the execution, and at the same time they bring to

the command a greater necessity of knowing, of foreseeing and of solving the difficulties of an attack; of undertaking none that cannot be carried through to the end, that cannot for that purpose be organized and begun under cover, protected till the last moment.

(7) The 2nd Battalion of the 37th Prussians which alone has borne a great part of the Austrian effort, and which has held the enemy all morning, has used up 23,000 cartridges; that is an average of 23 per man. Considerable results can be obtained, therefore, with even a small use of ammunition, on condition that fire be at all times well controlled,

VIII

STRATEGIC SURPRISE

THE idea of protection, which we have found to be of such importance in tactics, to rule always the leadership of troops, whether it be a question of information and safety for a maneuver, of preparing and assembling the means of action or of applying them to a known objective, that idea appears again in the forefront of the considerations on which strategic dispositions must be based.

Where there is no strategic protection, there may be strategic surprise, that is a possibility for the enemy of attacking us when we are in no condition to meet him, or the possibility for him of preventing our assembly of insufficiently protected troops. Meanwhile our forces wander and risk themselves in wrong directions through the insufficiency of the information obtained.

As a historical example of what I refer to, we will take the

STRATEGIC SURPRISE OF AUGUST 16TH, 1870
(See Map No. 6)

On August 15th there was, in the German 1st Army:

The 1st Corps at Courcelles-Chaussy;

The 7th Corps between Pange and the station of Courcelles;

The 8th Corps at Orny.

The 2nd Army, on the same morning, was continuing its advance towards the Moselle, with:
The 3rd Corps to reach the Seille at Cheminot;
The 12th Corps to reach it at Nomeny;
The 9th Corps remaining at Peltre;
The 10th Corps coming to Pont-à-Mousson, whence it was to send detachments towards the north, into the Moselle valley and on the plateau to the west.

The Guard was assembling at Dieulouard, and the 4th Corps marching towards the Moselle at Custines, its advance guard on Marbache.

On receipt of the news of the battle of Borny, fought on the evening of the 14th, these dispositions were slightly changed about 7 A.M. The 3rd Army Corps was to halt at once, the 12th was to occupy positions between Solgne and Delme, both corps to be prepared to support the 9th Army Corps left at Peltre, in case the French should assume the offensive on the right bank of the Moselle.

Then the advance was resumed, after the receipt of a telegram, sent by General Von Moltke from the Flainville height at 11 A.M., reading:

"The French have been thrown back completely on Metz, and it is probable that by now they are already in full retreat on Verdun. The three corps on the right (3rd, 9th and 12th) are now at the entire disposal of the Commander-in-Chief. The 12th is already on its way to Nomeny."

Finally, in the evening, the 2nd Army was disposed as follows:

3rd Army Corps { 5th Division, Novéant; advance guard, Gorze. 6th Division, Champey.

10th Army Corps		Pont-à-Mousson, 19th Division at Thiaucourt.
12th " "		Nomeny.
9th " "		Verny.
Guard		Dieulouard, advance guard at Quatre-Vents.
4th Army Corps		Custines, advance guard at Marbache.
2nd " "		Han-sur-Nied.
5th Cavalry Division	{	Xonville, Puxieux and Suzemont, in contact with the French.
Cavalry Division of the Guard	{	Brigade of Dragoons, Thiaucourt. Brigade of Cuirassiers, Bernécourt. Brigade of Uhlans, Ménil-la-Tour.

These dispositions were in accordance with the views of the German General Staff which, since the battle of Spickeren, thought it need no longer consider any French army on the right bank of the Moselle.

The pursuit of the French had been intrusted until now to the 1st Army, and the Germans had simply massed, especially on the morning of the 15th, the right of the 2nd Army in order to support the 1st Army if necessary, at the time when the French seemed inclined to battle.

The disposition of the 2nd Army on the march, with a width of 28 to 30 kilometers, of a greater depth, requiring one day at least for concentration, was a perfectly safe disposition so long as the enemy refrained from any action, which was not quite certain.

As early, therefore, as the 14th, General Von Moltke had wished to make the situation clearer. In an order of that date, for the purpose of ascertaining definitely the situation of his opponents, he had instructed the 2nd Army to move *all* its cavalry to the left bank of the Moselle towards the French lines of communication between Metz and Verdun, supporting it towards Gorze and Thiaucourt by the corps which would first cross the Moselle.

In accordance with that order, the 5th Cavalry Division had reached, on the 15th, the neighborhood of Xonville and Puxieux, while the 10th Corps had advanced one of its infantry divisions to Thiaucourt.

But, instead of *all* the cavalry, the 2nd Army had only detailed the 5th Division to the reconnaissance ordered; the 6th Division was still, on the 15th, at Coin-sur-Seille, and on the 16th it was to get entangled into the 3rd Corps; the Division of the Guard was dispersed over the Thiaucourt—Bernécourt—Ménil-la-Tour front, rendering any action impossible, especially against the still distant roads from Metz to Verdun; the Saxon Division of the 12th Corps remained with its army corps.

On the evening of the 14th, the 1st Army has fought long and hard. The battle, begun in quite an unexpected manner, has been directed in an improvised fashion by the Germans. It might have been fatal to them, as General Von Moltke admits, if the French had had the idea, as they had the means, of pushing back the heads of the German columns which were drawing too close.

On August 15th, at 11 A.M., General Von Moltke telegraphs to the commander of the 2nd Army that the French have been thrown back completely on Metz, and are probably already in full retreat on Verdun. On the other hand, Frederick-Charles shows us by the records of the General Staff that:

"The information sent by General Headquarters during the day of the 15th, together with various reports, and particularly those of the 3rd Corps, had led the commander of the 2nd Army to believe that the French army was hastily retreating towards the Meuse, and that he must therefore follow it without delay.

" A telegram sent as early as 11 A.M. of the 15th had informed General Headquarters of that opinion, and of the intention to cross the Moselle on the 16th with the army's main body. No counter-order having been received, Prince Frederick-Charles had issued, at 7 P.M., the following orders for the 16th:

" Pont-à-Mousson, August 15th, 7 P.M.

" During the last evening, parts of the 1st Army have attacked the enemy before Metz, and thrust him back into that place.

" The French army has begun its retreat towards the Meuse. Beginning to-morrow, the 2nd Army will follow the opponent in the direction of that river.

" The 3rd Corps will cross the Moselle below Pont-à-Mousson, to reach, via Gorze and Novéant, the main road of Metz-Verdun, either at Mars-la-Tour or at Vionville; its Headquarters will arrange to establish itself at Mars-la-Tour.

" The 6th Cavalry Division can move in advance via Prény and Thiaucourt to reach from Pagny the road mentioned.

" The 10th Corps which, preceded by the 5th Cavalry Division, is already marching on Thiaucourt, will continue to-morrow towards the Verdun road to about Saint-

Hilaire and Maizeray, bringing up as rapidly as possible the detachments still remaining behind, in Pont-à-Mousson and in the valley of the Moselle.

"The 12th Corps will leave Nomeny, and mass in Pont-à-Mousson, pushing an advance guard as far as Regniéville-en-Haye.

"Its cavalry division will advance towards the Meuse.

"The Guard will have an advance guard at Rambucourt, its main body and Headquarters near Bernécourt.

"The 4th Corps will push its advance guard to Jaillon, Headquarters at Saizerais.

"Efforts will be made to open communications, towards Nancy, with the 3rd Army.

"The 9th Corps will reach Sillegny, cross on the day after to-morrow the Moselle at Novéant, over the bridge thrown by the 3rd Corps, and will follow that corps towards Gorze.

"The 2nd Corps will bring the head of its column to Buchy.

"The cavalry divisions which precede the army must, at the same time as they advance, reconnoiter the outlets and crossings of the Meuse in view of their future employment by the 10th, 3rd and 9th Corps at Dieuze and Génicourt; by the 12th at Bannoncourt; by the Guard, the 4th and the 2nd at Saint-Mihiel, Pont-sur-Meuse and Commercy."

Meanwhile, the commanders of the 1st and 2nd Armies had received from General Headquarters the following instructions, dated from Herny at 6.30 P.M.

"So long as there is no certainty concerning the number of enemy forces left in Metz, the 1st Army will maintain, in the neighborhood of Courcelles, a corps which must be relieved as promptly as possible by the troops coming from Sarrelouis, under command of General Kummer. The two other corps of the 1st Army will take up positions, on the 16th, along the Arry-Pommerieux line, between Seille and Moselle. A bridge will immediately be thrown over the latter river, if the 3rd Corps has not already done so. Information is expected very shortly as to the movements of the 2nd Army on the 15th; concerning the new dispositions to be taken, they must rest on the following considerations:

"The advantage obtained, last evening, by the 1st and 7th Corps, together with parts of the 18th Division, was achieved under circumstances which preclude all idea of pursuing it. Only by a vigorous offensive of the 2nd Army against the Metz-Verdun roads, via Fresnes and Etain, can we gather the fruit of that victory. The Commander-in-Chief of the 2nd Army is instructed to carry out the operation in accordance with his own inspiration, and with the help of all the means at his disposal.

"The heads of the 3rd Army's columns have reached to-day the Nancy-Dombasle-Bayon line; its cavalry is beating the country towards Toul and the south.

"The General Headquarters of His Majesty will be established at Pont-à-Mousson to-morrow, beginning at 5 P.M."

We shall see later the results obtained from these decisions. But, in order to discuss them intelligently we must also examine the basis on which they rest, and the operations they will necessitate.

As to General Von Moltke's first message at 11 A.M.

of the 15th that "The French have been thrown back completely on Metz, . . ." it is true that at 11 A.M. on the 15th French columns have been seen retreating along the whole front towards Metz. But to say that the French have been thrown back completely is to exaggerate the facts accomplished, and therefore what effect they may have.

Into the battle of the 14th, the Germans have only engaged the approximate equivalent of 3 divisions. The battle has lasted five hours, from 3 P.M. to 8 P.M. One has brought back no trophy; only few prisoners have been captured; no ground has been won.

Under these circumstances, and with such poor tactical results, one could not describe the French as completely thrown back on Metz, nor consider their main body defeated. If they retired, therefore, it is because they were ordered to do so, and not because they had been compelled. The results of the 14th might be enormous from a strategic point of view, but they were nil from the tactical point of view. When the French army were met again, therefore, one must expect to meet not a beaten army, but troops in full possession of their material and moral resources.

As a deduction from this already exaggerated description of facts, Von Moltke added: "It is probable that by now they are already *in full retreat* on Verdun. . . ."

That probability, foreseen at General Headquarters and resulting only in the return to the army commander of the three corps previously immobilized, would seem unlikely to have unfortunate results.

But the *possibility* foreseen, the *probability* communicated to the Pont-à-Mousson Headquarters, becomes there a *certainty* under the influence particularly of reports from the 3rd Corps, as we have seen. The commander of the 2nd Army then *decides* that the French Army is " hastily retreating towards the Meuse."

Those reports of the 3rd Corps only announced, however, some French movements from Metz to Longeville. That was still far from the Meuse.

As to the 5th Cavalry Division, it could have supplied highly important information if it had been consulted; it would certainly have said that the roads of Mars-la-Tour and of Conflans were free of the French, except near Vionville, where cavalry bivouacs were to be seen.

The conclusion of Prince Frederick-Charles seems, therefore, a preconceived idea; the Prince so readily joins into the hypothesis of General Von Moltke that for him it requires no proof.

To make assurance doubly sure, however, he transmits this view at 11 o'clock to General Headquarters. The Chief of Staff, a cautious and reserved man when he has no certain information, does not answer all day. He knows the value of that hypothesis: " It is probable that the French are already in full retreat on Verdun." One can substitute for it another one, just as sound, that of Prince Frederick-Charles to the effect that " the French army is hastily retreating towards the Meuse." He therefore allows the commander of the 2nd Army to act in accordance with his opinion and march to the Meuse, intervening only at 6.30 to say:

" The advantage obtained on the 14th by the 1st Army was achieved under circumstances which preclude all idea of pursuing it. Only by a vigorous offensive of the 2nd Army against the Metz-Verdun roads can we gather the fruit of that victory. The Commander-in-Chief of the 2nd Army is instructed to carry out the operation in accordance with his own inspiration, and with the help of all the means at his disposal."

We note two things in that order:

(1) It is a *victory* which is described as needing to be taken advantage of, whereas one knew, or should have known, that it was nothing of the kind.

(2) The fruit of that victory is to be gathered by a vigorous offensive against the Metz-Verdun roads.

Such an objective is already much more reasonable than the Meuse, because, closer to the source of the truth, Von Moltke is less inclined to exaggerate.

Such an objective can be attained, he believes, by the 2nd Army. He well knows that, owing to its distribution, it cannot reach there as a whole, that it cannot, on the 16th, employ there all its forces; it needs twenty-four hours at least to concentrate without advancing, and if it crosses the Moselle more than forty-eight hours will be required to bring up the most distant corps (on the 18th, one of them, the 4th, will still be lacking). Thus its dispersion will bring against 3 French armies, whose morale has been improved by the resistance of the 14th, 1 German army only, the 2nd, much reduced because of its dispersion.

Von Moltke realizes the insufficiency of his combination, and so ends his order by leaving to the commander of the 2nd Army the decision as to how the operation is to be carried out, "*in accordance with his own inspiration.*"

He is reduced to inspirations. It cannot be otherwise,

because Von Moltke, lacking reliable information, has based his whole combination on hypotheses. He doubts himself their worth.

For lack of a plan depending on protection which alone insures safety, there can only be inspiration, more or less lucky.

Von Moltke believes no more in his own than in that of Prince Frederick-Charles. It is only in a way as an indication that he communicates his guess to the latter, and what maneuver he would employ in connection with it; but he realizes that he can impose neither one nor the other. He therefore leaves the commander of the 2nd Army free to act in accordance with his own inspiration, and to employ what maneuver he pleases, with all the means at his disposal, in spite of the known impossibility for part of the 2nd Army's forces to operate on the 16th along the left bank of the Moselle. Having opened the door to error, he ceases to effectively command.

Let us now examine the order of Prince Frederick-Charles. It is dated from Pont-à-Mousson, 7 P.M.

The commander of the 2nd Army will abandon himself to his inspiration, as he is invited to do, and as there is no criticism of his manner of interpreting the facts. Inspiration thus continues to be the basis of the German strategic combinations.

Being further from the sources of truth, he takes the hypothesis of Von Moltke for a certainty, as we have seen, and a certainty well advanced.

"The French army is hastily retreating towards the Meuse," and therefore a vigorous offensive against the Metz-Verdun roads becomes useless. A vigorous offensive is not needed against a defeated army which is hastily retreating. By aiming at the Metz-Verdun roads one would miss the objective. It is to the Meuse that one must hasten. And so, the 2nd Army will at once follow the enemy towards that river.

As we have already seen, if Prince Frederick-Charles had sought to confirm his guess by information sought from the cavalry (5th Division) on the roads leading to Verdun and to the Meuse, the 5th Division would have answered that on the 15th there were no French columns on the roads, except a cavalry division at Rezonville; that consequently the French army could not be there before the 16th; that it would take more than one day for its columns to pass a given point, Mars-la-Tour for instance; it had 5 corps (2nd, 3rd, 4th, 6th and Guard) and only two available roads. But the Prince, taking his *inspiration* as sole basis for his decision, neglecting *information* which would have revealed the truth, neglects also thought and reasoning to employ only imagination. He aims therefore at the Meuse. Besides, as the French army is hypothetically beaten, and as the Meuse is distant, it is a race which he plans in order to reach there the enemy columns. The army will be dispersed in order to move quickly, speed being the sole object.

Von Moltke's already insufficient combination which consists in attacking the French army while in fact that army is still intact, with only one of his three armies, and that an incomplete one, weakens still in passing by the hands of Prince Frederick-Charles. For the latter, instead of planning an attack, organizes a pursuit; instead of an army seeking battle, it is a dispersed army which he will present: a disposition full of weakness to deal with a French army that has suffered no important reverse, and whose morale has been raised by the victorious resistance of the 14th.

Moreover, this point of view which nothing justifies, and the maneuver resulting from it, run counter to the point of view of Von Moltke, not imposed of course, yet clearly indicated in the words: " Only by a vigorous offensive. . . ." Prince Frederick-Charles will seek to simultaneously accomplish the march to the Meuse and the offensive against the Metz-Verdun roads, being unafraid to take two objectives at once. The dispersion of forces, necessitated already by the idea of pursuit, will be made greater still.

Of any combination of forces there is no trace. Thus, while seeking these two objectives, so different and so distant from one another, we end in the order of August 15th by the maximum employment of the roads available, and the unlimited dispersion of forces, in respect to front and in respect to depth.

As regards the cavalry divisions which precede the column, they need not, according to the order, seek the enemy to lead those columns to him, nor need they discover him in order to discover possible danger. They simply have to reconnoiter the crossings of the Meuse.

On the 15th, decisions of the highest importance have been taken from inspirations resting on no certain information; the army has been all dispersed without maintaining, through a system of protection, the possibility of concentrating it again if necessary.

On the 16th, inspiration and confidence still reign supreme; without any information or protection, orders are given for reaching the enemy, and no thought is even spared on the time and space necessary to concentrate the army if this dispersed enemy should suddenly reappear.

The uncertainty of the blows planned is equalled only by the hazardous position deliberately adopted.

The idea of protection, allowing mastery over the unknown and the danger of surprise in war, is quite lacking in the strategy of the 2nd Army. Von Moltke has realized its advantage when he ordered, on the 14th, to send on the left bank of the Moselle all the cavalry of the 2nd Army, and to support it with infantry. But he did not realize its absolute necessity.

In consequence of these orders, and in case the enemy did not attack, the 2nd German Army was to be, on the evening of the 16th, in the following situation:

    3rd Corps: Vionville, Mars-la-Tour;
    10th   "   Saint Hilaire, Maizeray;
    12th   "   Pont-à-Mousson;
    Guard   Bernécourt;
    4th Corps: Saizerais;
    9th   "   Sillegny;
    2nd   "   Buchy.

As to the other two armies, the 1st and 3rd, neither would be able to render any assistance for several days.

Such dispositions only allowed the *pursuit of a beaten enemy,* concerning whose retirement there would have been *certain information,* and if these conditions were not realized, there was the greatest possibility of danger. For if the enemy appeared, either to take the offensive, or merely to carry out a retirement concerning which there were no exact details available, the German army had deliberately rendered itself incapable of meeting him;

the forces of even one army could not be assembled before meeting with a serious check. *Surprise* was complete.

The error by which Von Moltke and Frederick-Charles altered, or at least exaggerated, the meaning of the battle of the 14th, had evidently assumed two different forms in accordance with the temperament of each of these two men.

Von Moltke is a Chief of Staff relying constantly on his intelligence, acting by reason, a thinker more than an executive. He determines the unknown by a hypothesis, logical enough but sprung merely from his imagination, which even he does not consider above doubt; he thus reaches a solution which he does not compel. Examining the various combinations which the enemy might adopt, he has chosen the most likely one, and it becomes the idea from which will spring his plan of maneuver. It is a probable one, but it is not going to be the true one. His lack of assurance in the merit of his reasoning will keep him from imposing it; he will advise, but not order, remaining a Chief of Staff and not a Commander of Armies. Hence the big results of the war were only in part due to him. Such he will appear at Sedan, where he again fails to command on August 30th, where the envelopment will result from an agreement of the two armies and not from a higher decision. Such he will be seen also during the operations on the Loire.

Frederick-Charles, on the other hand, is essentially a man of action; the mere thought of a possible great result intoxicates him to the point of depriving him of the power to appreciate its starting point or to realize fully its extent and the dangers it entails. He changes the hypothesis of Moltke into a certainty. He starts recklessly, and until the end he will remain blind.

He seeks, as we have seen, no information on the 16th; but, worse still, he dictates on that day at noon an order covering the arrival on the 17th of the whole 2nd Army on the Meuse. He still relies on the so-called victory of the 14th in this order which the records of the General Staff have carefully preserved although no part of it has been carried out. As a document, it is the strongest and bitterest criticism of the dispositions taken by the Prince on that day:

" General Headquarters of Pont-à-Mousson,
" August 16th, 1870; noon.
" Army Order

"The 2nd Army will continue to-morrow its advance towards the Meuse. The 1st Army will very shortly come behind its right flank. In consequence of the direction taken by the enemy in his retreat, that flank of the 2nd Army will carry out its movement as follows:

" The 10th Corps will ultimately cross the Meuse below Verdun. It will detach troops towards that place.

" If the pursuit should result in drawing this army corps far enough north, the positions of Clermont-en-Argonne and Sainte-Menehould are already designated for its direction, to then become the right flank of the army.

" The 3rd Corps will advance to-morrow on Etain, where it will throw out an advance guard, unless the situation as regards the enemy causes it to decide otherwise. The troops left to guard the bridge thrown over the

Moselle will join up as soon as the 9th Corps has had them relieved, which is to be done to-day.

"The 9th Corps will reach, to-morrow also, Mars-la-Tour. If it be possible, this army corps will, during the same day, replace the bridge thrown by the 3rd Corps by another, built from boats seized on the Moselle; after which, the Bridging Section of the 3rd Corps will proceed to rejoin. The three corps of the right flank named above will keep up communications, and every day they will notify their position to my General Headquarters, at the places named below. In case of a serious encounter with the enemy, Infantry General Voigts-Rhetz is authorized to dispose of the 3rd Corps first, later of the 9th also if it be necessary.

"If, as is to be expected, there is no such encounter, the 3rd and 9th Corps will move on the 18th, the former towards Dieue-sur-Meuse, the latter towards Fresnes-Génicourt in order to seize as promptly as possible the bridges across the Meuse.

"In case the 9th Corps should arrive ahead of time, it would guard simultaneously both these means of crossing.

"The 12th Corps will bring, to-morrow, the head of its column to Vigneulles, its main body to Saint-Benoit-en-Woëvre, where its Headquarters will be located. The cavalry will be thrown on, and across, the Meuse.

"On the 18th, this corps will continue on Bannoncourt, and seize the outlet there on the Meuse.

"The Guard will march to-morrow on Saint-Mihiel, sending a strong advance guard on the left bank of the Meuse in order to protect that important crossing, and establishing its Headquarters at Saint-Mihiel.

"The Guard Cavalry will move in advance towards Bar-le-Duc.

"The 4th Corps will shortly advance on Commercy by the Jaillon-Sauzey-Boucq line, unless the position of Toul should necessitate a partial halt in the movement.

"The 2nd Corps will reach Pont-à-Mousson to-morrow, and establish the head of its column in the direction Limey-Flirey-Saint-Mihiel. Headquarters at Pont-à-Mousson.

"My General Headquarters will be to-day at Thiaucourt, after 5 P.M.; to-morrow, afternoon, it will be located until further orders at Saint-Mihiel.

"After the 2nd Army reaches the Meuse, and the bridges on that river being guarded, the troops will probably be stationary for a few days, until the armies on our flanks reach the same line.

"Every army corps will send daily a Staff Officer to my General Headquarters. These officers may, if necessary, use carriages, followed by their saddle-horse, and have an escort of infantry orderlies.

"The General Officer Commanding the Cavalry,
"*Signed:* Frederick-Charles."

When, after the war, Von Moltke discusses the decisions of the 15th, he allows but little justification to Prince Frederick-Charles. Thus he writes:

"The 3rd and 10th Corps, together with the two cavalry divisions attached to them, were intended to make a strong demonstration against the Verdun road (strong, except that it covered only two army corps separated by more than 15 kilometers, and liable to meet the French army between Saint Hilaire and Mars-la-Tour).

STRATEGIC SURPRISE 271

" As to the other portions of the army (4 corps) they maintained a purely westerly direction towards the Meuse.

". . . The plan of the commander of the 2nd Army was therefore intended to direct the result of operations towards the Meuse. If the French were not found on the Moselle, one hoped, thanks to the good marching qualities of the German troops, to join them on the Meuse.

" The information received from the 5th Cavalry Division during the day of the 15th had not clearly shown the true conditions. (The 5th Division had not been asked for facts, because they had been guessed at. Otherwise, the cavalry would have advised that the French were not retiring over the roads from Metz to Verdun; reasoning would also have shown that their movement must last 48 hours at least; I have already pointed this out.)

" The instructions of General Headquarters, received in Pont-à-Mousson at 10 P.M. of the 15th attached, it is true, special importance to the occupation of the roads from Metz to Verdun; but by sending in that direction two army corps and two cavalry divisions one was justified in believing that the recommendation had been sufficiently complied with."

On the morning of the 16th, therefore, the several parts of the German army started in accordance with orders, but the quiet assurance which reigns in the mind of Prince Frederick-Charles is not shared by all those under him, as Von Moltke also tells us:

"Infantry General Voigts-Rhetz (commanding the 10th Corps), worried by those French bivouacs reported on the previous day, considered it necessary to combine, with the movement of his army corps on Saint-Hilaire, a strong reconnaissance on the encampments observed during the evening of the 15th near Rezonville. He had assigned to that operation the 5th Cavalry Division under command of General Rheinbaben, reinforcing it on the morning of the 16th by two batteries of horse artillery, escorted by the 2nd Squadron of the 2nd Regiment of Dragoons of the Guard. In order to support that reconnaissance, an order was also given to half of the 37th Infantry Brigade, then at Thiaucourt, to rejoin at Chambley a detachment of Colonel Lynker sent to Novéant, in the Moselle valley. General Voigts-Rhetz intended to march, meanwhile, from Thiaucourt to Saint-Hilaire with the remainder of the 5th Division."

Thus the higher command has decided that no protection was needed, but a subaltern commander realizes the danger. He can only, however, obtain an imperfect form of protection, and too late to repair all the misfortunes which may have occurred. It is evident, therefore, that the advance guard should have been organized from the very start.

From that investigation made by the 5th Cavalry Division, soon continued by the 6th Division arriving on the heels of the 3rd Corps, some knowledge of the situation will at last be obtained.

We know what they discover:

The 3rd Corps has put its foot on the ant-hill. The French army, instead of being in full retreat towards the Meuse, finishes the evacuation of Metz; instead of being beaten, it is of excellent morale. It is massed between the two roads of Conflans and Mars-la-Tour, 6 kilometers from Gorze. Against these forces appears the 3rd Corps.

At 11 A.M., when the battle is well started, all the army corps other than the 3rd are on their way to reach the appointed destinations:

The 10th by the Thiaucourt-Saint-Benoit-Maizeray road, at an average distance of 19 kilometers from Vionville;

The Guard at double the distance, about 40 kilometers;

The 4th at treble the distance, 55 kilometers;

The 12th, 9th and 2nd, in second line, are more than a day's march further back.

Under these circumstances, the 2nd Army can only present to the French forces debouching from Metz:

On the 16th, one army corps, the 3rd, and the greater part of the 10th;

On the 17th, three to four army corps.

It must await the 18th to assemble the greater part of its forces.

That is a true strategic surprise in the fullest sense of the word. It consists in the brutally sudden appearance of considerable enemy forces when the enemy presence is not suspected so near, for lack of information, and when one cannot mass, for lack of protection.

The movement has been deliberately undertaken into the unknown as the result of a hypothesis (Von Moltke), of a conviction without reason or verification (Frederick-Charles).

Let us therefore conclude that:

(1) In strategy as in everything else, a leap into the unknown is highly dangerous; we have no right to substitute, for facts ascertainable, the creations of imagina-

tion. From facts alone can a logical maneuver be planned.

(2) In strategy as in tactics, no maneuver must be undertaken which causes dispersion, unless we have first assured for ourselves the possibility of being able to regroup our units when necessary.

IX

STRATEGIC SAFETY

IF we go back to an instance taken from the beginning of the campaign of the Army of Italy in Hungary during 1809, we shall find there an explanation by Napoleon to Prince Eugene of the method which aims at developing a strategic maneuver with *certainty* and *safety*.

It is the end of May, 1809. The Army of Italy (Prince Eugene) in pursuit of Archduke Jean has reached Gratz. It has moved towards Leoben, thence towards Vienna, to carry out at Brück on May 26th its junction with the Army of Germany, while Archduke Jean retires from Gratz into Hungary, for the purpose of either maneuvering on Raab or operating from there against the still dispersed forces of the Army of Italy.

Macdonald arrives, via Laybach, in Gratz on the 30th. He remains there, awaiting orders from Prince Eugene and news of Marmont, who is coming independently from Istria and Carniola. (*See Map No. 7.*)

The general situation, as regards the French, shows therefore three divisions at Neustadt, the Macdonald column arriving in Gratz, Marmont's Corps marching in the same direction; the whole is protected by a strong advance guard under command of Lauriston, pushed by the Army of Germany at Oedenburg.

Napoleon writes to Eugene:

"Ebersdorf, June 3, 1809, 10 P.M.

"My son, General Lauriston advises that the advance guard of Prince Jean seems to be moving on Oedenburg, or at least that instead of passing via Körmend he has moved between Körmend and Oedenburg via Rechnitz. This would presume that his Corps intends to assemble towards Raab, and that because he proposes to follow the road of Körmend he has himself protected on the left to 7 or 8 leagues from Körmend. It is not impossible even that, having learnt from the inhabitants of the small numbers at Oedenburg, he may wish to strike a blow at that city. I see no objection to your moving your Headquarters to Oedenburg (General Grouchy can proceed there from Brück without going through Neustadt), and to your pursuing Prince Jean in order to cut off his retreat, on condition of merely taking care that he does not pass to your right, that is between you and Brück, or between Oedenburg and Neustadt. . . .

"I leave you free to move on Oedenberg without giving you any definite order, because I presume you have information from your right which enables you to act in accordance with my intentions, which are shown by this idea: *that you seek to harm Prince Jean.*

"You can do so if he retires on Raab; you can do nothing, without movements so important as to lead you away from the army, if he retires on Pesth. Finally, in Oedenburg you will be no further away from the army than from Neustadt. You must know what there is at Friedberg and at Hartberg."

Prince Eugene marches on Oedenburg, where he seeks first of all, by sending towards Körmend his cavalry supported by one battalion, to obtain information concerning that opponent who appears restless, and might later either move to the Danube and thence join Archduke Charles, or by Körmend act against Macdonald before the junction of all the French forces coming from Italy. Napoleon writes him:

"Ebersdorf, June 5th, noon.
"My son, I receive your letter of the 4th, 9 P.M. I approve the move you have made on Körmend. But the cavalry should not have gone there without the infantry. I fear that this battalion so far from the forces may be compromised. As it appears from General Macdonald's letter that the enemy is still opposite Wildon, and that the Gyulai Corps is towards Radkersburg, a strong cavalry detachment on Körmend, pushed in the rear of the enemy, might protect our communications, especially if it be supported by a strong detachment from General Macdonald to the same point. Write to him in that sense. General Macdonald must not send a reconnaissance, but a strong advance guard to Fürstenfeld, and thence to Körmend."

Napoleon does not therefore believe that the cavalry alone can fulfill the first part of the task; he asks that the cavalry be supported at Körmend by a *strong advance guard* coming from Fürstenfeld. It is from a body so strongly organized, and not from a number of squadrons, that he expects a real reconnaissance.

But meanwhile, at Oedenburg Prince Eugene has conceived the idea of moving on Raab to cut off Archduke Jean from the Danube road, by which he might join Archduke Charles. That is a maneuver resting on a preconceived idea; it may fail, or it may at least be parried. We shall see Napoleon's opinion:

"Schoenbrunn, June 6th, 1809, 9 A.M.
"My son, I receive your letter of the 5th, 10 P.M., and see that Colbert has finally met Archduke Jean. The first thing you must do is to advance together and united. I do not consider that the Seras and Durutte Divisions, and the five cavalry regiments of General Grouchy, are sufficient. It is necessary that the Corps of Baraguey d'Hilliers and the Guard be with you, so that you have in hand 30,000 men marching united, capable of acting all together and of reaching the same battlefield within three hours' time.

"I leave at your disposal the Corps of Lauriston, which will reinforce you by 3,000 infantrymen, and the three cavalry regiments of Colbert. I also leave at your disposal the Montbrun Division, which consists of four cavalry regiments. By that means, you will have eleven regiments of light cavalry, three regiments of Dragoons and a corps of nearly 36,000 men. Send at least half of these 36,000 men as advance guard to march on Körmend. The Duke of Auerstaedt is opposite Presbourg with the Gudin Division and the Light Cavalry Division of General Lasalle. You will not receive this letter before noon; it is impossible that you should not have news by then from General Lauriston, from General Montbrun, from General Colbert and even from General Macdonald, giving you a clear idea of the situation of Prince Jean.

"In such plains as those of Hungary, one must maneuver differently than in the mountains of Carinthia and Styria. In the gorges of Carinthia and Styria, if one moves ahead of the enemy to a point of intersection, such as Saint-Michel for example, one cuts off an enemy column; but in Hungary, on the other hand, the enemy, as soon as he loses the race at one point, will move to another. Let us suppose that the enemy is moving on Raab, and that you reach this town ahead of him: the enemy learning of it on his way, will change his direction and move on Pesth; . . . then your movement on Raab would take you further from him, and might even give him the idea (for the enemy is not like us; being at home, he is well informed) of attacking and defeating Macdonald. I think, therefore, that the movement first on Güns, later on Stein-am-Anger, later on Körmend, or from Güns on Sarvar, is the wisest movement, if however you have no other information than I now have here. This evening, you can march on Güns with the brigade of Colbert, the seven regiments of the Grouchy Division and much artillery (you must place your light artillery, at least twelve guns, with your cavalry), and the Seras and Durutte Divisions. The corps of Baragney d'Hilliers can arrive this evening at Oedenburg, or even reach Güns, or march to the intersection of the road from Sarvar and Raab to Zinkendorf. According to the information you receive, you can continue to-morrow the movement of your two columns on Sarvar or on Körmend. General Montbrun must have been last evening, 5th, at Gols, and as he is to connect with General Lauriston, you will not fail to have news."

From that letter we note that in a country of easy communications like Hungary (and the same would be true of a great part of Europe), the enemy keeps his freedom of movement so long as we have not immobilized him. The maneuver on Raab may therefore:

(1) Either fail if he does not come there;
(2) Or be parried: losing the race to that point, he moves to another;
(3) Or even cause a crisis: incite the enemy to attack and defeat Macdonald.

The wisest maneuver is to march on Güns, later on Stein-am-Anger, later on Körmend, in the direction where the enemy has been *observed*.

How to march there? With a strong advance guard followed by a united main body.

Taking up the question again on the next day, Napoleon writes:

"Schoenbrunn, June 7th, 2.30 A.M.

". . . In your pursuit of Prince Jean from the Tagliamento, you have not advanced with sufficient unity, and we might have suffered. If Prince Jean had assembled his forces at Tarvis, it was possible that you would be unable to beat him. You were split up into 3 bodies. Macdonald, Seras and yourself. . . . You will understand that I make these remarks for your advantage. You must advance with everybody well united, not in little bundles. Here is the general principle in war: a corps of 25,000 to 30,000 men may be isolated; under good leadership, it can fight or avoid battle and maneuver according to circumstances without disaster, because it cannot be forced into an engagement and because it can fight for a long time. A division of 9,000 to 12,000 men can be left, without danger, isolated for an hour; it will hold the enemy, however numerous, and give time to the army to arrive.

It is not usual, therefore, to form an advance guard of less than 9,000 men, and its infantry must be kept well together, an hour's distance from the army. You have lost the 35th because you have ignored that principle; you have organized a rear guard of only one regiment which was turned. If there had been four regiments, they would have formed such a mass of resistance that the army would have arrived in good time to their assistance. . . .

". . . To-day you will start on a prepared operation; you must move with an advance guard composed of much cavalry, of a dozen guns and of a good infantry division.

"All the remainder of your corps must bivouac one hour behind, the light cavalry protecting of course as much as possible. . . .

"From the advance guard to the tail-end of your transport, there must not be over three or four leagues. . . ."

And so, the strongly organized advance guard with which we are familiar is not only the reconnaissance unit which we have had to send from Oedenburg to Körmend. It is also the mass of resistance such that the army may arrive in time to reinforce the advance guard and continue the business undertaken, in order to strike the enemy held at last.

In short, inversely from Prince Eugene who organizes a service of information and a maneuver independent of one another, resting on the nature of the ground and the supposed intentions of the enemy, Napoleon conceives the maneuver as a prolongation of the information obtained, changing therefore in accordance with circumstances, thanks to the powerful advance guard capable of:

(1) Supporting the searching parties in quest of news;

(2) The enemy being found, undertaking for its own account the service of information, and for that purpose changing search into reconnaissance;

(3) When the enemy has been found and his strength ascertained, immobilizing him during the time necessary to the army's arrival.

The army's main body follows behind, ready to take immediate advantage of these results.

We shall see later how these ideas might have been applied to the situation of August 15, 1870, which has been discussed already.

For the moment, let us remember Napoleon explaining to Prince Eugene strategic safety, along the following lines:

". . . You knew nothing of the enemy's intentions, you were quite right in sending towards Körmend all your cavalry, in seeking information. But you were wrong to send it without infantry, because it cannot by itself suffice to the task; it is with a strong advance guard that you should have sent it. . . .

"You were wrong also in seeking to proceed at once to Raab in order to cut off Archduke Jean's road. He will know of your movement and escape you. . . ." In countries with easy communications one cannot determine in advance the maneuvers against an enemy free of his movements. First one must seize him, and once that is done, one can maneuver with certainty and safety.

The advance guard which has carried out the first duty, information, must then also attend to the second, holding the enemy to allow of a reasoned and true maneuver in accordance with the circumstances. For that purpose it attacks the enemy if he seeks to escape. It resists by defensive and withdrawal if he attacks.

To undertake operations implies therefore, in the Emperor's mind, to march on the enemy with an advance guard and main body capable:

One, the main body, of carrying out a maneuver hastily planned in accordance with circumstances;

The other, the advance guard, of insuring safety, that is of supplying reliable information from which to plan the maneuver, and safety during its preparation and initial execution.

This idea of safety will be found by us throughout Napoleon's wars, with various dispositions according to the moment and the operation undertaken, but always based on a combination of time, space and force.

The idea of strategic safety is quite lacking, on the other hand, in the German armies of 1870, and that fault has often placed them in very critical situations. It took all our lack of initiative to allow them to extricate themselves sometimes without disaster.

But strategic safety was known and applied by the Germans of 1813 and 1814. Instructed by the painful lessons from the Emperor, they have realized its importance. They even jeer at such French generals as ignored its necessity. Thus, Clausewitz wrote:

"Has one not seen, in spite of the methods of Emperor Napoleon, some French corps of 60,000 to 70,000 men advancing under Marshal Macdonald in Silesia and under Marshals Oudinot and Ney in the Marche, without any question of an advance guard!"

He refers by these words to our defeats of Katzbach, Donnewitz and Gross-Beeren.

EXAMPLE OF DISPOSITION OF STRATEGIC SAFETY
(*See Map No.* 8)

The beginnings of the campaign of 1815 show us clearly how, in the Army of the Lower Rhine, one knew how to organize and employ advance guard troops. When the enemy assumes the offensive, it is the concentration of forces which must first of all be carried out. The 1st Prussian Corps shows us the tactics to be employed for that purpose by the general advance guard.

During the first half of June, 1815, the army was distributed as follows:

1st Corps
Gen. Ziethen
H. Q.: Charleroi
- 1st Brigade at Fontaine-Lévêque;
- 2nd " " Marchiennes;
- 3rd " " Fleurus;
- 4th " " Moustier-sur-Sambre;
- Reserve cavalry at Sombreffe;
- Artillery at Gembloux.

2nd Corps
Gen. Pirch
H. Q.: Namur
- 5th Brigade at Namur;
- 6th " " Thoremberg-les-Béguines;
- 7th " " Héron;
- 8th " " Huy;
- Reserve cavalry at Hanut;
- Artillery Along the Louvain road.

3rd Corps
Thielmann
H. Q.: Cinay
- 9th Brigade at Assesse;
- 10th " " Cinay;
- 11th " " Dinant;
- 12th " " Huy;
- Reserve cavalry Between Cinay and Dinant;
- Artillery at Cinay.

4th Corps Bülow H. Qu.: Liège	13th Brigade at 14th " " 15th " " 16th " "	Liège; Warenne; Voiroux-Gorey; Voiroux-les-Liers.
Reserve cavalry	1st Brigade at 2nd " " 3rd " "	Tongres; Dahlem; Lootz.
Reserve artillery	Gloms; Aihem.	

Altogether, the forces amounted to 110,000 fighting men.

That disposition of the army was far from perfect from the military point of view. It was chiefly the result of the existing difficulty to provision the troops. Blücher received no money from his government; the authorities of the country were little disposed to supply him with any; he must depend on the inhabitants to feed the army during a long period of time. One did not expect to take the offensive before July 1st, and the troops had been in the country since the month of May. Because of this circumstance, it had been necessary to spread them out.

Offensive was still intended, however, by the Prussians; but, while awaiting the arrival of the third allied army (Austrians, Bavarians, Wurtemburgers) on the theater of operations, they had complied with Wellington's request to delay large-scale offensives, and, in case of unexpected attack from Napoleon, it had been agreed that the two Prussian and English armies would carry out their junction on the road from Namur to Nivelle via Sombreffe.

In fact, to an opponent like Napoleon who "dared stake the whole decision in the great action of a unique battle eagerly sought" (Clausewitz), one could reply only by a concentration of the two allied armies, at the proper time, on the same spot or on two spots so near to one another that they might act simultaneously. Tactics would then bring victory from the great numerical superiority obtained through this first operation.

In accordance with these ideas, "the Prussian Army occupies with two army corps the valley of the Meuse, where the cities of Liège, Huy and Namur supply quarters for numerous troops. It has one corps (the 1st) on the Sambre towards Charleroi; another (the 3rd) towards Cinay on the right bank of the Meuse, extended forward like antennae; General Headquarters are in Namur, central point, 3 or 4 miles from the corps in front, and joined to Brussels by a main road. It covers a space of 8 miles in width and 8 miles in depth; it can therefore gather within two days on its center, and it is assured of two days for that purpose. Once assembled, it can either accept battle, if it considers itself strong enough, or withdraw in any direction, for it has in its neighborhood nothing to keep it back or to limit its freedom of action.

"At Blücher's Headquarters, one had chosen as point of assembly for the Prussian army the ground of Sombreffe. The stream of Ligny and a small tributary form, on a parallel with the Sombreffe-Saint-Balâtre road, a cut in the ground which is neither very steep nor very deep, although sufficiently so to provide on the left slope of the valley an excellent position for the employment of all arms. It had an average stretch (one-half mile), so that if occupied by one or two corps it could keep up a long resistance. Blücher retained therefore two corps for an offensive, and could thereby decide the fate of the battle, either alone or with Wellington."

In the Prussian general's mind, the idea of concentration, and the point where it must be carried out are therefore clearly established. There remains to be considered the possibility of carrying it out at a suitable time.

"The point of Charleroi is the nearest to the place of concentration; it is only three-and-a-half miles distant. If the news of the enemy's arrival comes from Charleroi to Namur, and thence the order of concentration to Liège which is the furthest place occupied, one must figure about sixteen hours before it reaches there, and eight hours more to notify and call out the troops; a total of twenty-four hours before the 4th Corps can start on its march.

"From Liège to Sombreffe there are two days' good marching; it will be a matter of three days, therefore, before the 4th Corps arrives at Sombreffe. The 3rd Corps at Cinay could arrive in thirty-six hours, the 2nd Corps at Namur in twelve hours.

"On the other hand, the resistance of General Ziethen on the Sambre, and his retreat to the neighborhood of Fleurus could not allow more than one day, from morning till night, to stop the enemy; the coming of night allowed the balance of the twenty-four hours.

"One could hope, moreover, that the enemy's route would be known before the first cannon shot; at the latest, when he took his last positions before assaulting the Prussian troops, and very probably also, from other information, a few days sooner.

"In the latter case, the time sufficed for assembling.

"If visible evidence only were available, if the enemy's intentions were known only by his attack on the outposts, the 2nd and 3rd Corps alone could reach the neighborhood of Sombreffe to assist the first; the 3rd would do so with difficulty, and the 4th would miss the concentration.

"This danger was clearly realized at Blücher's Headquarters, but many difficulties stood in the way of joining Bülow's Corps, especially the question of provisions. As soon, however, as any movement was noticed on the French side, on the 14th, it received the order to march towards Hanut, only 5 miles from the point of concentration, where it would be nearer to the 3rd Corps at Cinay.

"Blücher believed he could assemble his army within thirty-six hours near Sombreffe. Although the chances were a hundred to one that the enemy's advance would be learnt more than thirty-six hours before his arrival in the neighborhood of Sombreffe, it was still very dangerous to remain so dispersed, with an advance guard so close at hand (the one at Charleroi). It would not have been done but for the constant difficulties of approvisioning due to local conditions; concentration would have been carried out first."

Such, explained by Clausewitz, is the whole theory of the time and space necessary to the operation, which time and space the advance guard is expected to provide.

Let us come to the facts, to see the actual operation of forces thus organized.

Napoleon, intending to begin operations on June 15th, moves forward:

On June 6th, the 4th Corps from Metz, and a few days before the 1st Corps from Lille; he masks these

When Foch stopped the Germans. This picture was taken after the most important council of the war for the Entente.

departures behind a reinforcement of the advance guards by territorial troops;

On June 8th, the Guard from Paris;
" " " the 6th Corps from Laon;
" " " the 2nd Corps from Valenciennes;

On the 12th, he left Paris himself.

All these corps reached, on the 13th, the region between Philippeville and Avesnes; on the 14th they assembled, and formed three columns:

On the right, the 4th Corps and cavalry;

In the center, the 3rd Corps, 6th Corps, Guards and the greater part of the cavalry, near Beaumont;

On the left, the 1st and 2nd Corps, near Solre-sur-Sambre.

These movements are quite unnoticed till the 14th, on which day the enemy learns of the Emperor's arrival at the army and of the French movements to concentrate; he still ignores where this concentration is taking place. At this first warning, Blücher orders, on the evening of the 14th, his 4th Corps to immediately assemble its troops so as to reach Hanut by one day's march.

Only during the night of the 14th-15th is the whole truth learnt, from information sent by General Ziethen: he sees the enemy being reinforced before him, and foresees that he will be attacked on the next day. At this second warning, Blücher follows the order already sent to General Bülow with another order to immediately proceed to Hanut.

This second order reached General Bülow on the 15th, at 11 A.M. If he had immediately ordered his troops to resume, after a brief halt, their march on Hanut, the 4th Corps would have been assembled there during the night of the 15th-16th. General Bülow decides that the carrying-out of the order can be put off until the next day, 16th. He moreover reported accordingly. But his report did not find Blücher at Namur, just as the orders sent on the 15th by Blücher had not found Bülow at Hanut, although he had been ordered to be there on the evening of the 15th. These orders instructed him to continue, on the 16th, the advance with his army corps from Hanut on Sombreffe. In reality, if he had carried out the orders received, the 4th Corps could reach Hanut in the night of the 15th-16th; from there to Sombreffe the distance is 38 kilometers; he could, on the 16th, reach it with his advance guard toward noon, and with the remainder of his troops in the evening. It would have been a forced march, it is true, but it would have enabled him to take part in the Battle of Ligny and perhaps change its outcome.

The 3rd Corps at Cinay also only received its marching orders on the 15th, at 10 A.M.; it reached, however, the battlefield on the 16th, about 10 A.M.; the 2nd Corps had arrived without difficulty.

With a better exchange of communications, the concentration of the four Prussian Army Corps would have been carried out in time, even with the dispersion necessitated by the special problems of provisioning the troops. In any case, it brought to the battlefield, on the 16th, three corps out of four, that is forces greater than those of the Emperor.

The result was due to the use of an advance guard, the 1st Army Corps, capable of securing the time and space needed for the operation planned: concentration.

It is through fighting a retreating engagement that this army corps succeeded, moreover, in lasting twenty-four hours without being destroyed by far superior enemy forces.

Retreating engagement, two words which seem to contradict one another; from which we find that the more one must retreat, the less engagement there is; and the less one retreats, the greater the engagements. The army corps could only retreat a short distance, 15 kilometers; it would have to fight several times, and we shall see how it succeeded.

The order which it had received from Blücher, dated Namur the 14th, 11 P.M., ordered it to *retire on Fleurus, in case it should have to deal with superior forces; however, it must not lose sight of the enemy, but energetically dispute the ground.*

In the early morning of the 16th, its forces were distributed as follows:

The 1st Brigade (Steinmetz) at Fontaine-l'Evêque, holding Thuin;

The 2nd Brigade (Pirch) at Marchiennes, with 2 battalions on outposts;

The 3rd Brigade (Jagow) at Fleurus;

The 4th Brigade (Henkel) at Moustier-sur-Sambre.

The outpost line, in the part which concerns us, ran through Thuin, Ham-sur-Heure and Gerpinnes.

As we know, the French army had been ordered to move in three columns:

On the left (1st and 2nd Corps) via Thuin and Marchiennes;

In the center (3rd and 6th Corps, Guard and cavalry

reserve) via Ham-sur-Heure, Jamioux, Marcinelle and Charleroi;

On the right (4th Corps) via Florenne, Gerpinnes and Le Châtelet.

About 4 A.M., the Prussian outposts were attacked; first of all those of the 2nd Brigade of the 1st Corps. It was the light cavalry of General Domon, leading the center column, which appeared first. The picket company at Ham-sur-Heure, vigorously attacked and surrounded by the French cavalry, was compelled to surrender; three other companies from the same regiment assembled at Gerpinnes, and succeeded in retiring from there into the small valley from Gerpinnes to Le Châtelet.

About the same time, Thuin was attacked by the French column on the left. Two battalions, five squadrons and three guns begin the action against this locality, occupied as we have seen by a German battalion. After about an hour's fight, the Westphalians, who have delayed at Thuin, are enveloped there. They seek to open up a way at the bayonet by the plateau of Montigny, and two squadrons of Prussian Dragoons try to support them. These squadrons are soon thrust back by the French cavalry; the battalion is partly cut to pieces, partly captured. Its destruction is due entirely to its delay in abandoning Thuin and in the direction it chose for its retirement. By following the slopes of the Meuse valley, which offer better cover, it could have avoided more easily the attacks of the French cavalry.

While all this was happening, General Ziethen, informed during the night of the attack which threatened him, had got all his troops under arms, and had ordered them to stand firm while awaiting information from the outposts. He did not intend by any means to resist with his main body on the outpost line. He only expected his outposts to inform him of the extent of any attack, but that maneuver shows one difficulty, that of withdrawing the outposts. The cavalry is instructed to support them: the 1st Regiment of Dragoons primarily; it soon needs to be reinforced by other squadrons.

Between 6 and 7 A.M., Ziethen has received reports which show the whole French army to be in movement. Especially is his 2nd Brigade threatened. It is ordered to avoid any important engagement, and with that object to establish a line of resistance on the Sambre, the crossings of which it must hold at Charleroi, Châtelet and Marchiennes, retiring eventually on Gilly.

The outpost line of the 1st Brigade has not been attacked, except at Thuin where it had one battalion; in spite of that fact, the Brigade is ordered to retire. It is to move on Gosselies, keeping in line with the 2nd to guard against envelopment.

The 3rd and 4th Brigades, the reserve cavalry and the army corps' artillery will assemble and occupy positions at Fleurus. We shall see later what use is made of them. They form a reserve which will be drawn upon to assist the troops engaged against the enemy.

These troops will successively and along the whole front organize resistance to compel the enemy to make dispositions for attack. When the attack thus prepared occurs, they will without much fighting abandon the ground, and take up further the same dispositions along a new line of resistance previously held by the echelons in the rear, towards Gosselies for the 1st Brigade, and towards Gilly for the 2nd.

At 8 A.M., the French cavalry reached the river after thrusting back all the outposts on the right bank of the Sambre. Coming from Marcinelles under command of Pajol, it appeared before Charleroi. A dam, and further a bridge, joined the village to the town. The bridge was barricaded. The French cavalry (4th and 9th Regiments of Chasseurs), mistress of Marcinelles, seeks to reach the dam and the bridge, but it is thrown back by the fire of the German snipers. Later, between 9 and 10 o'clock, the attack is resumed by the 1st Hussars, who try an assault of the bridge; they are stopped by violent fire from the barricade. To force that position, infantry is needed. Pajol decides to wait. About 11 o'clock, the Emperor himself arrives with sailors and sappers of the Guard, bringing also the " Jeune Garde." Sappers and sailors rush the bridge and open the way to Pajol's cavalry. It ascends at a smart trot the steep street which runs through Charleroi from north to south.

The Prussian battalion of Charleroi had already withdrawn; in good order, it marches to the position of Gilly; the charges of the French cavalry have no effect on it.

While the French center column was thus attacking towards Charleroi, the left column was attacking Marchiennes. In accordance with the Emperor's intentions, it should have occupied Marchiennes at 9 o'clock. But the stubborn resistance of the battalion at Thuin has delayed the movement. Almost two hours are also needed to prepare the attack of the bridge of Marchiennes. In fact, it is nearly noon before the bridge is carried, when Charleroi has already fallen.

The movement of retreat on Gilly of the 2nd Prussian Brigade entailed, in accordance with orders given,

STRATEGIC SAFETY

that of the 1st Brigade towards Gosselies. In order to facilitate that retirement of the 1st Brigade, Ziethen had detached during the morning some troops at Gosselies, namely the 29th Infantry Regiment (of the 3rd Brigade) and the 6th Uhlans (of the army corps' cavalry reserve). Shortly after noon, these troops were in position, with one battalion at Gosselies and two in reserve further back, while the 1st Brigade began the crossing of the Picton stream. It was the moment when the French, debouching from Charleroi, started the pursuit: towards Gilly with the Pajol cavalry soon followed by the " Jeune Garde"; towards Gosselies with the 1st Hussars under Clary.

Colonel Clary, having reached Jamet, attacks Gosselies; he is repulsed by the 20th Regiment, while the 1st Prussian Brigade, thanks to the resistance of the 29th Infantry, completes the passage of the Picton stream and reaches Gosselies. As soon as the 1st Brigade has passed through the Gosselies defile, the 29th Regiment retires on Ransart. The 1st Brigade, instead of continuing its movement on that locality, seeks to hold up the enemy on the road to Gosselies. It is promptly attacked and thrust back by Colonel Clary, who has been reinforced by the advance guard of the 2nd French Corps, coming from Marchiennes; the brigade is also cut off from Ransart, which the Girard Division of the same army corps has just carried; it retires on Heppignies, protected by the 6th Uhlans and 1st Hussars.

In these happenings around Gosselies clearly appear the difficulties of such movements of retreat, and also the means of overcoming them.

The chief danger lies in being cut off from the line of retreat by enveloping movements of the enemy. The remedy is found in supports (29th Infantry, 6th Uhlans) placed in the rear to receive the retreating force (1st Brigade).

The facts show the use made of such supports. They stop the Clary cavalry until the 1st Brigade has extricated itself from its dangerous position. That service being rendered, the supports immediately withdraw to occupy the important points on the line of retreat (Ransart and Ransart Wood). The retreating force should have followed that movement without any delay. It cannot think of stopping, by battle, forces far superior to its own (Clary cavalry reinforced by the 2nd French Corps). That is not the mission with which it is intrusted; on the other hand, it risks being destroyed, or being cut off from its line of retreat by these superior forces. That is what happens. Steinmetz, arrived at Gosselies with his 1st Brigade, discards the assistance (29th Infantry) held out to him, which has saved him once before. Instead of continuing his retreat on Ransart, which is held by the 29th, he stands still at Gosselies and prepares for a battle in which he is defeated. He is compelled to resume his retreat on Heppignies, but he is now cut off from the army corps by the French Division of Girard which has seized Ransart; he will have trouble in joining his corps.

Events of a similar nature were occurring at the same time on the road of Fleurus and Sombreffe through Gilly.

It is in Gilly that General Pirch had received instructions to assemble his brigade, to carry out a second recon-

STRATEGIC SAFETY

naissance, the one on the line of the Sambre having been broken.

We have already seen how he had gradually abandoned the positions of Marchiennes and Charleroi, decreasing his holdings on the Sambre as the enemy columns came up. In the same order of ideas, he withdrew from Le Châtelet the 28th Infantry which occupied it first, and substituted for it the 1st Western Prussia Dragoons. Thus he had assembled at Gilly the greater part of his brigade at the time that the French entered Charleroi. There only remained to withdraw the detachments of Marchiennes and of Charleroi.

Pirch placed his brigade behind Gilly, its front protected by the muddy stream of Grand-Rieux. Four battalions and the brigade's battery are on the slope of the stream's left bank, namely:

The 2nd of the 28th to the north of the road, covered by the Abbaye de Soleilmont;

The F of the 1st to the south of the road, against a small wood;

The F of the 28th behind and to the left of the above;

Behind, and to the right of the F of the 1st, the artillery (4 guns) on a low height south of the road; two other guns between this position and the road; two more north of the road commanding the entrance to Gilly;

The 2nd Battalion Westphalian Landwehr in reserve, behind the artillery;

Three more battalions are in reserve near the road to Lambusart (1st of the 28th, 1st and 2nd of the 1st Regiment).

The direction of retreat of the Brigade was on Lambusart. In order to protect it against a sweeping attack which the French would not fail to carry out through Gilly and the Fleurus road, that road had been covered with abatis cut from the trees.

The occupation of the position thus obtained was guarded on the left, at Le Châtelet, by the 1st Western Prussia Dragoons, intrusted also with the maintenance of communications with the brigade which had occupied Farciennes; on the right, protection was supplied by a mounted picket (1 officer and 30 cavalrymen) at Ransart, which General Steinmetz was to occupy. We know by what tactical blunder General Steinmetz was cut off from Ransart, thus upsetting these carefully prepared plans.

That situation of the Prussian advance guard will remain unchanged all afternoon, until 6 P.M.

Pajol, debouching from Charleroi about noon, had marched on Gilly, where he was soon followed by the detachment of the Exelmans cavalry; the arrival of all that cavalry, combined under the orders of Grouchy, resulted in a complete evacuation by the Prussians of the immediate surroundings of Gilly, where they had protected till the last possible moment the main position north of the stream. Grouchy was unable to attack that position with nothing but cavalry. He returned to report the situation to Napoleon at Charleroi, while the " Jeune Garde " arrived before Gilly, and while the Vandamme column reached Charleroi. It was after 3 o'clock.

Napoleon was mounting his horse to make a personal examination of the situation. In spite of the length of the Prussian line, knowing moreover the dispersed condition which a surprise attack always encounters on the side of the enemy, he did not expect to find more than ten thousand men opposite him. He therefore ordered a frontal

STRATEGIC SAFETY

attack with one of the Vandamme divisions, supported by Pajol's Division of Cavalry, while Grouchy with the Exelmans Division was to maneuver on the enemy's left flank, which offered the best opportunities.

An enveloping maneuver is especially appropriate against a rear guard, because the latter no longer fulfills its mission as soon as it is turned.

These orders having been given, Napoleon returns to Charleroi for the purpose of watching what occurs on the Gosselies road, and of apparently hastening the advance of the Vandamme Corps. His absence brought back indecision to the minds of the French generals before Gilly; they thought they were opposed by considerable forces. They knew of movements carried out by troops of the 3rd (Jagow) Brigade, arrived from Sombreffe. It must be reinforcements that had arrived, they thought. These reports and the difficulty of reconnoitering through the wood delayed their action. They took over two hours to combine their attack. About 5.30, Napoleon, worried at hearing no guns from the direction of Gilly, hastens up again. He doubts neither the state of surprise nor the state of dispersion of the enemy, and he does not doubt especially the need for speed on his own part. He orders the attack; it is almost 6 P.M. A French battery of 16 guns opens up the attack. Then, all preparations having been made behind the height of the windmill, near the Grand-Trieu farm, three columns debouch; the one on the right moves towards the little wood occupied by the F of the 1st; the center one leaves Gilly on its left and advances on the center of the position; the one on the left passes to the north of that village. They are supported by the Pajol cavalry.

The Prussian battery has soon suffered heavy losses under the French artillery's fire. Skirmishers from both sides are engaged, when General Ziethen orders General Pirch to retire. Hardly have the Prussian battalions begun the movement before they are charged by the French cavalry. It is the Emperor who has ordered General Letort to charge with the squadrons at hand. Seeing that the Prussians may reach the woods and thus escape him, he throws against them any cavalry available. Letort does not even take time to group his four squadrons; he starts off with those of the 15th Dragoons, leaving the others to follow when they can. Crossing the stream north of the road, and then the road itself before the columns of Vandamme, he strikes the retreating Prussians. The F of the 28th are hit first, and lose two-thirds of their numbers; then the F of the 1st who are still 500 meters away from the wood, and have had time to form a square and open fire. The charge of the French cavalry has already slowed up, and that battalion succeeds in reaching the wood, along the fringe of which it posts one company to stop all pursuit.

General Letort had been mortally wounded. At the same time, the Exelmans Division, debouching above the Châtelet, had bowled over the 1st Western Prussia Dragoons and put to flight a reserve battalion holding the Pironchamps wood. To these cavalry attacks was added one from the Pajol Division, succeeding at last in passing the columns of Vandamme. The whole Pirch Brigade was retreating on Lambusart, where it endeavored vainly to reorganize. The French cavalry left it no time to do so, and it fell back on Fleurus, and later on Sombreffe. The attack was discontinued, night having fallen. The

STRATEGIC SAFETY

Vandamme Corps went into bivouacs between Winage and the Soleilmont wood, protected by all Grouchy's cavalry before Fleurus.

On the Brussels road, Marshal Ney had halted the heads of his columns in line with Gosselies, sending on to Mellet only one (Bachelu) division and the light cavalry of Piré, and detaching to Quatre-Bras the Lancers and Chasseurs of the Guard. At the end of the day, his most advanced troops were in Froesnes. The Prussian Steinmetz (1st) Brigade had regained by a detour the road to Sombreffe.

The losses suffered by the Prussians, at the time of their undertaking the retreat of Gilly, clearly show the difficulties which troops encounter in disengaging themselves from an attack when they delay too much the beginning of the movement. The need to do so arises sooner in modern times because of the greater effective range of newer weapons.

The next morning, 16th, Grouchy informed the Emperor from Fleurus that strong columns, apparently coming from Namur, were moving towards Brye and Saint-Amand, behind Fleurus. They consisted of the 2nd and 3rd Corps rejoining the 1st. In spite of the absence of the 4th Corps, the Prussians would be able to dispose that day of nearly 90,000 men, considerably more than the Emperor had.

The Corps of Ziethen had suffered considerable losses, but obtained the very important result of retarding the battle till the 16th, thereby permitting concentration.

As Clausewitz says: " One sees thereby what caution

and what slowness are inevitably imposed by circumstances of but a slightly complicated nature even on the most resolute of Generals, Napoleon."

Among the complications of which Ziethen took advantage, we must include the double retreat on the Gilly and Gosselies roads, which prevents Ney's going to Quatre-Bras, making Napoleon's intervention there necessary and, in that way also, delaying action on the Namur road.

Observe, moreover, that this double retreat does not prevent the 1st Prussian Army Corps from grouping its four divisions the next day.

That example shows how advance guards fight when retiring, inspired by the double duty of observing the enemy and delaying his approach.

They delay the enemy by compelling him to assume battle formations, to assemble and deploy in order to use his superiority for an enveloping movement.

The nature of the ground and the distance from the body protected determine of course the length of the resistance; and from the resistance offered will depend the losses sustained. For that reason, rear guard actions are to be avoided when the necessary time can be obtained in another way.

Normally and logically, we must only seek therefore to hold the enemy and to delay him by three methods:

(1) By enforcing prudence from him, and therefore slowness of advance;

(2) By prolonging as long as caution permits, but never later, local resistance;

(3) By retreating as slowly as possible.

That retreat, as slow and deliberate as possible, must allow the troops to reorganize and take up new positions along the road. It is necessary, therefore, that the local resistance and the movement of retreat prolong one another, and that the struggle only cease at a given point when it can be repeated, with equal method, at a number of other points.

An advance guard can delay the enemy in proportion to its own strength; the opponent, to enforce its retreat, will need more time to develop the necessary means of doing so.

"It is less by their actual actions than by the mere fact of their presence, less by fighting than by constantly threatening to fight, that advance guards fulfill their mission. They do not prevent the enemy's action, but, like a pendulum, they moderate and govern its movements, and thus permit of ascertaining its mechanism and extent" (Clausewitz).

We have seen the difficulties of a rear guard action:

(1) Danger of being turned: once turned, the protective troops no longer guard the main body; they may, moreover, be cut off;

(2) Danger of letting the enemy come to close quarters, which makes it very difficult to disengage all the troops;

(3) Necessity of battle by fire and at long range, to act on the enemy from afar.

The employment of troops that fulfill these various conditions generally necessitates the occupation of each successive position by a large proportion of artillery, usually all that is available. A sufficiently large quantity allow the troops to reorganize and take up new positions lery, while the remainder of the infantry prepares and carries out the occupation of the second position.

A good deal of cavalry is also needed to detect and parry enveloping movements. It generally constitutes the reserve at every position occupied.

In this manner, protective troops consisting of 6 battalions, 6 batteries and 6 squadrons will generally employ on the first position their 6 batteries, 2 to 3 battalions and their 6 squadrons, while the other battalions organize the second line of resistance, where the artillery will join them at the trot when it abandons the first line. The cavalry finally protects the retreat of the last infantry units from the first position, and afterwards resumes its function of general reserve.

To protective troops maneuvering in retreat to guard the main body, just as to protective troops advancing to seek and hold the enemy, much cavalry is therefore necessary, supported by artillery and infantry.

The proportion of the various arms to be used varies, however, with the distance from the main body. An advance guard at a short distance will increase its means of resistance (infantry and artillery), and decrease its cavalry; because information at close range is of little value, and because the space available allows little maneuvering in retreat, compelling battle therefore. It is what happens with the 1st Prussian Corps in 1815.

In either case, if the maneuver be ready when the enemy appears, the battle begins, and the protective troops are reinforced as much as possible by troops capable of prolonged action: masses of artillery. Pro-

STRATEGIC SAFETY

tected by that first disposition, the maneuver for the battle begins.

STRATEGIC PROTECTION APPLIED TO THE SITUATION OF AUGUST 15, 1870

The 2nd German Army on August 15th, 1870, needed, as we have seen, forty-eight hours to concentrate when advancing, moving to the left bank of the Moselle. How could it, by applying sound theories, have avoided the crisis which occurred?

On the morning of August 14th, 1870, the French army in its retreat has arrived by Metz: the day's battle will change nothing in that situation; it may only delay the retreat. On the other hand, the main body of the French army has not been beaten, and one must therefore beware of it.

In any case, beginning on this 14th day, a direct pursuit of the French becomes impossible, owing to the protection given to them by the position of Metz.

To attack again, one must act on the left bank of the Moselle, crossing first of all that river in presence of an enemy who can, during several days more, attack along either bank.

The crossing itself is dangerous enough to make it necessary that, putting aside all thought of future action towards the Meuse, or towards the roads, or towards Metz, one should devote to it all possible protection.

What is known of the enemy? That he was near Metz on the 14th. If it was desired to transport by the 16th the main body of the 2nd Army to the left bank, the operation could be in accordance with the following dispositions:

(1) Locate the army's crossing at, and above Pont-à-Mousson, at a distance allowing the army's concentration before it could be seriously attacked by the enemy.

(2) Move the troops on the 14th and 15th towards these crossings, protected by an advance guard on the river's right, in contact with the enemy; combine the advance and dispersion of the forces with the resistance of the advance guard, so that, if the enemy appeared from Metz on either the 14th or 15th, the whole army might accept battle on the right bank within twenty-four hours.

(3) The enemy having failed to attack, thrust by surprise, during the morning of the 16th, the army on the left bank of the Moselle, under protection of a new advance guard on that same left bank insuring time and space for the operation, for assembling the troops and for engaging them, even if the French attacked on that 16th of August.

In accordance with that idea, three roads could be used, and allotted as follows to the 3rd Army:
Cheminot, Pont-à-Mousson 3rd and 9th Corps;
Nomeny, south of Pont-à-Mousson, Blénod (1 bridge to build) 10th, 12th and 2nd Corps;
Lixières, Dieulouard Guard and 4th Corps.

The 1st Army would have kept its function as advance guard on the right bank, with defensive duty therefore, for the day of the 14th; the corps of the 2nd Army would on that day have lengthened their march as much as possible, and would have brought their heads:

Those of the first line near the Moselle, namely:

3rd at Cheminot; 12th at Atton; Guard at Dieulouard; the whole 10th at Pont-à-Mousson.

Those of the second line having their heads just behind the tails of the above.

All the forces were thus on the right bank of the Moselle, in a position that allowed them to concentrate there for battle within twenty-four hours. Any attack coming from the French would first have struck the 1st Army; the latter would have been held or would have maneuvered in retreat, according to the degree of concentration of the 2nd Army. Battle was made possible with both armies.

On the 15th, the 1st Army, still acting as advance guard, would have reached the neighborhood of Fleury, Chesny and Courcelles.

The units of the 2nd Army would have closed up their intervals.

During the evening of the same day, the first line troops, which had marched but a short distance, resumed the movement about 11 P.M. to cross the Moselle, an operation accomplished by 6 A.M. of the 16th.

The second line units, beginning their march at 5 A.M. on the 16th, towards the Moselle, would also have crossed the river by noon. The whole army would then be, at noon of the 16th, on the left bank.

But to insure having the time and space to carry out that movement in safety, an advance guard would have been sent to the left bank on the evening of the 15th.

For that purpose, the 10th Corps, concentrated on the evening of the 14th at Pont-à-Mousson, would have picked up arms in that city the 15th at noon and, reinforced by the army's four cavalry divisions, would have proceeded to Chambley, to act from there as advance guard on the left bank. It could have insured the possibility for the 2nd Army of assembling as a whole on the left bank during the morning of the 16th.

Moreover, the necessity of an advance guard on the right bank gradually ceased as the 2nd Army carried out its crossing. On the very evening of the 15th, as a matter of fact, the advance guard (1st Army) at Fleury, Chesny and Courcelles was no longer needed. That army, which had marched little during the 14th and 15th, might have resumed its movement on the evening of the 15th towards the Moselle, beginning with its second line troops. It would have crossed the river below Pont-à-Mousson, near Pagny, over bridges specially built for it, protection being supplied by the 10th Corps at Chambley.

On the morning of the 16th, it could have two corps on the left bank, the third being kept on the right bank or drawn back later without difficulty.

The 10th Corps, reaching Chambley on the evening of the 15th, would naturally have sent troops (important pickets of infantry, artillery and cavalry) to occupy all the dangerous roads at Gorze, at Vionville, at Mars-la-Tour. The main body of the cavalry, kept near Xonville and Sponville, would have provided good sources of information on the front and further north, towards the road leading from Metz.

The main body of the army corps, assembled in the region of Buxières, Chambley, Hagéville, Saint-Julien, etc., would have left a rear guard and its train on the Rupt-de-Mad towards Rembercourt.

By acting thus, one kept conditions safe on the right bank during the 14th and 15th. As to the movement prepared for crossing the Moselle, the enemy could not notice it before the evening of the 15th. What dispositions he planned on the evening of the 15th to interfere with that crossing could be realized only during the morning of the 16th, too late to keep the army from crossing the river and concentrating if necessary.

Should the enemy have made any plans before the evening of the 15th, the movement of the 10th Corps would reveal them, and allowed of overcoming them or of meeting them with counter-dispositions. In any case, it would have pointed out to the army the danger which threatened it, and avoided *surprise*.

X

THE BATTLE: DECISIVE ATTACK

IN our studies of tactics, we have come across infantry in battle, cavalry in battle and artillery in battle. But from the glimpse we have had of various forms of action, we may have failed to deduce the logical reasons for them all.

The battle must never be considered like a drama that succeeds through careful staging, a plot cleverly carried out or the characteristics of the persons represented. Nor is it the development of a maneuver, a methodical and successive employment of the various arms, having all proportionate shares in the ultimate result.

But the battle can be described in no such way. Far from being a total of distinct and partial results, it is the one result of many efforts, some of them successful, others apparent failures, aiming all at one goal: the decision which alone gives victory. Either there must be a successful ending or the whole effort has been wasted: " In war, as long as there remains something to be done, nothing is accomplished," said Frederick. Every move in the battle must therefore work to that end. And, inasmuch as there is direction, combination and results, it proves that logic rules the actions, with all its privileges and pitiless severity. There is a theory of the battle.

We shall therefore study that *conclusion*, victory, in

THE BATTLE: DECISIVE ATTACK

itself, also the *method* of finding it, and we shall then have learnt the general principles which must inspire the decisions of the higher command, as well as the actions of every leader of whatever importance, so that the work on hand may be well carried out.

Modern war, in order to reach its purpose: to impose one's will on the enemy, knows of but one means: the destruction of the opponent's organized forces.

That destruction is undertaken and prepared by battle, defeating the enemy, disorganizing his command, his discipline, tactical unity, his troops as forces.

It is realized by pursuit, in which the victor, profiting by the moral superiority which victory gives him over the vanquished, cuts up finally troops that have become demoralized, dispersed, impossible to command, troops which are no longer troops.

Such was not the case with the engagements which we have studied in the preceding chapters, advance guard, rear guard or flank guard actions, at Nachod for instance. All had only a limited purpose, determined in every case, preparing for the battle but not being the battle, although they all necessitated a considerable use of force. At Nachod, the problem was for the 5th Prussian Corps to debouch from a defile and open its doors to the 6th Corps behind; the task of the Austrians was to prevent this. Some similar problem occurs in every engagement which arises from the service of protection. In every case the tactics to be practiced depend on the particular characteristics of the aim in view, and of the time and place available.

To-day we deal with the battle, the sole argument in

war, sole object therefore of our strategic operations; and we wish to find out whether there are any tactics of conquest, and *what* they are.

Let us establish first that, in order to duly fulfill the double purpose of being the logical aim of strategic operations and the effective means of tactics, battle cannot be merely defensive.

Under that form it may, it is true, halt the enemy in his advance; it keeps him from attaining some immediate objective; but such results are purely negative. Never will it destroy the enemy or procure the conquest of the ground he occupies, which is the visible sign of victory; it is unable, therefore, to ever create victory.

A battle of this kind, purely defensive, does not, even if well conducted, make a victor and a vanquished. It is merely something to be decided again later.

A purely defensive battle is like a duel in which one of the men does nothing but parry. He can never defeat his opponent, but on the contrary, and in spite of the greatest possible skill, he is bound to be hit sooner or later.

Hence we find that the *offensive* form, whether it be immediate or as succeeding the defensive, can alone give results. It must consequently be *always* adopted at some stage or other.

Every defensive action, then, must end by an offensive blow or successful counter-attack if any result is to be gained. It is an elementary principle, if you wish, but neglect of it has been frequent. It was not understood by the French Armies of 1870, or they would not have pictured as victories days like the 14th or 16th of August, 1870, and many others, which might have *be-*

THE BATTLE: DECISIVE ATTACK

come victories but which certainly were not victories at the stage where they were left. The French had merely held their positions, which is not synonymous of victory, and even implies future defeat if no future offensive action be undertaken.

" To make war was always to attack." (Frederick.)

We must always seek to create events, not merely to suffer them, we must first of all organize the attack, considering everything else of secondary importance and to be planned only in respect to the advantages which may result from it for the attack.

The necessity of the offensive form of battle having been admitted, is victory obtained from a number of conflicts of individuals or of small units? Will it come, rather, from an intelligent combination of forces? Napoleon tells us that " 2 Mameluks could hold out against 3 Frenchmen; 100 Frenchmen did not fear 100 Mameluks; 300 could defeat an equal number, and 1,000 would beat 1,500, so great was the influence of tactics, of discipline and of proper movements." Individual quality of the men is not sufficient, therefore, to create victory. Decisive as it is in single combat, it loses of its weight as numbers increase. If Napoleon had carried his explanation further, he would have reminded us that at the Battle of the Pyramids a handful of Frenchmen, commanded by himself, had defeated about thirty thousand of these valiant Orientals, individually superior to the French.

There must be therefore such a thing as wise tactics and intelligent battle formations, that is a combination of forces by leadership. The influence of the leadership, of the command, becomes considerable and decisive; it triumphs over individual quality when numbers increase.

Where shall we find the method whose existence is now evident? Will it consist in the number of enemies killed? Is it a question of doing more harm by having more guns and more rifles, or better guns and better rifles, than the enemy? Is superiority found merely in material advantages, or does it come from other causes? We must seek the answer in an analysis of the psychological phenomenon of battle.

"A hundred thousand men," says General Cardot, "leave ten thousand of their number on the ground and acknowledge defeat: they retreat before the victors who have lost just as many men, if not more. Besides, neither know, when the retreat occurs, what their losses are or what the enemy's casualties may be." It is not, therefore, through the material factor of losses, and still less through any comparison of figures, a greater number of casualties, that they give in, renouncing the fight and abandoning to the opponent the ground in dispute.

Ninety thousand defeated men withdraw before ninety thousand victorious men solely because they have had enough, and they have had enough because they no longer believe in victory, because they are demoralized and have no *moral* resistance left. Which leads Joseph de Maistre to say: "A battle lost is a battle one believes one has lost, for a battle is never lost materially." And if battles are lost morally, they must also be won in the same way, so that we can add: "*A battle won is a battle in which one refuses to acknowledge defeat.*"

THE BATTLE: DECISIVE ATTACK

De Brack, following Frederick, passes by an old castle in Silesia. Over the door is a coat-of-arms: two stags with locked horns, and for a motto: *The most obstinate wins*. "That is the truest source of success!" exclaims the famous general.

Proofs and instances could be given indefinitely of that great importance of morale in war. Von der Goltz himself tells us that: "It is not so much a question of destroying the enemy troops as of destroying their courage. Victory is yours as soon as you convince your opponent that his cause is lost." And again: "One defeats the enemy not by individual and complete annihilation, but by destroying his hopes of victory."

In order that our army be victorious, its morale must be higher than that of the enemy, or it must obtain such superiority of morale from the higher command. To organize the battle we must, therefore, in order to break the enemy's morale, raise ours to the highest pitch.

The will to conquer: such is the first condition of victory, consequently the first duty of every soldier; and it is also the supreme resolution with which the commander must fill the soul of his subordinates.

That necessitates, for an army that desires to conquer, the highest sort of command, and it necessitates in the man who undertakes to battle one important quality: the ability to command.

"It is not the Roman legions that conquered the Gauls, but Cæsar. Not the soldiers of Carthage caused Rome to tremble, but Hannibal. It was not the Macedonian troops that penetrated as far as India, but Alexander. During seven years Prussia was defended against the three most powerful nations of Europe, not by Prussian soldiers but by Frederick-the-Great."

Napoleon wrote these words, but he could have written more, and with still better cause, if he had included that wonderful period of history which he has completely filled with his own personality.

Great results in war are due to the commander, and it is justice that history couples with the names of famous generals victories that glorify them or defeats which dishonor their memory.

Again let me quote the words of Scharnhorst at the time of Blücher's appointment to command the army of Silesia in 1813: "*Is it not the manner in which the leaders carry out the task of command, of impressing their resolution in the hearts of others, that makes them warriors, far more than all other aptitudes or faculties which theory may expect of them?*"

The facts soon confirm that opinion of Blücher, who is still considered in court circles as an imbecile and sickly old man, while by his influence in the country—he represents in the eyes of his countrymen the idea of patriotism—he has won universal confidence and popularity. He has, moreover, won the complete devotion of his men by the constant care he takes of their welfare, and he will be able to demand everything, undertake everything and obtain everything. By his considerable influence, this man, of no brilliancy but possessed of a determination which is never discouraged, will lead whole nations to victory.

The necessity of such influence is easy to understand. On the field, when the time comes to make decisions, to incur responsibilities, to bear sacrifices, when initiative must be preserved everywhere, where can we find the

THE BATTLE: DECISIVE ATTACK

men we need unless it be among those of unusual determination and unusual greed for responsibility?

"One little realizes the strength of mind necessary to deliver, fully grasping its consequences, one of those battles from which depend the history of an army and of a country, the possession of a throne," says Napoleon. And: "by a strong mind we must not mean one that only knows strong emotions, but one which even the strongest emotions cannot sway."

Such a leader finds a way to increase the power of his forces, and he transforms the resources at his disposal, creating efficient subordinates and worthy troops, capabilities and devotions where, without the spark or impulsion from above, there would probably have been only banal mediocrity.

This immense task of the command can rarely be carried out, with the size of the present armies, by one man only. It requires a number of men, with subordinates using their own initiative towards a unique purpose.

But battles are not fought without reason: "Battles of which one cannot say *why* they were fought, and with what purpose, are the usual resource of ignorance."

Yet history shows us many such battles, as around Metz in August, 1870, when we find an army fighting bravely when its chief did not desire victory.

The great events of history, the disasters which appear on some of its pages, such as the collapse of French power in 1870, are never accidents. They can be traced to higher and general causes which are omissions of the most ordinary moral and intellectual truths. It is therefore necessary, if we wish to clearly understand war, that we recognize first its main principles.

How can an army efficiently commanded destroy the morale of the enemy? Into what actions is war, display of moral force, translated?

To answer that question, we need only see how a mental impression is created.

"Everything," says Xenophon, "pleasant or terrible, causes us the more pleasure or fear in proportion as we have least expected it. This is nowhere more evident than in war, where every *surprise* brings terror even to those who are most powerful."

He says *terror* "the cold goddess Fear, but not the fear of a woman who runs away screaming. That one we can, and even must, consider as an impossibility though it be not a quite unknown phenomenon, but that other and much more terrible fear which descends on the strongest heart, chills it and persuades it that it is defeated." (J. de Maistre.)

The way to destroy the enemy's morale, to show him that his cause is lost, is therefore surprise in every sense of the word, bringing into the struggle something "unexpected and terrible," which therefore has a great effect. It deprives the enemy of the power to reflect, and consequently to discuss.

It may be some new engine of war, possessed of novel powers of destruction, but that cannot be created at will; ambushes and attacks in the rear are suitable to small-scale warfare, but impractical in big operations where we must resort to the sudden appearance of a danger which the enemy has no time to avoid, or which he can only partly avoid. It may be the apparition of a destructive force greater than his own, necessitating a concentration of forces, and of overwhelming efforts, at a point where

THE BATTLE: DECISIVE ATTACK 319

the enemy is in no position to parry instantaneously by a similar deployment of forces within an equal time.

To surprise, is to crush at close range by numbers and within a limited time; otherwise, the enemy surprised by greater numbers is enabled to meet the attack, to bring up his reserves, and the assaulting forces lose the advantage of surprise.

They lose it also if the surprise begins from afar, for the enemy can, thanks to the range of weapons and to their delaying powers, regain time to bring up his reserves.

Such are the conditions of numbers, of time and of space which must be observed in order to obtain those characteristics of surprise which are necessary for the destruction of the enemy's morale.

Hence appears the superiority of maneuvering armies, alone capable of speed for:

Preparing an attack;

Beginning it at close range;

Carrying it through rapidly.

In the same manner we find the community of characteristics and effects realized in the flank attacks of our predecessors, in the oblique movements of Frederick or in the operations of Napoleon.

Under various forms we always find the same principle of *surprise,* seeking to produce on the enemy the same moral result, *terror;* creating in him, by the sudden appearance of something unexpected and overwhelming, the feeling of impotence, the assurance that he cannot win; that is, that he is beaten.

To destroy the enemy's morale is therefore the first

principle we find; to destroy it through an unexpected blow of overwhelming force, such is the first consequence of that principle.

But that overwhelming and unexpected blow need not be struck at the whole enemy army. To defeat an opponent, it is unnecessary to "simultaneously cut off his arms, his legs and his head while piercing his chest and stabbing him in the stomach." (General Cardot.) In the same way, to overcome an army's flank, its center, any important part of the whole, will be sufficient for the result sought.

The army, moreover, is a delicate being which thrives only on discipline. Discipline is the main strength of armies, but it is also the prime necessity to their existence; discipline alone, thanks to hierarchic organization permitting the transmission and execution of orders, enables the commander to control any action.

To break the chain is, therefore, to transform the tactical units into disjointed human masses; it renders the execution of orders impossible, it destroys the will of the commander, it prevents all action. And to break the chain, it is enough to spread moral or material disorder, to upset the organization at one point of the whole.

All this brings us to the application of one overwhelming blow at one point, which Napoleon has put into words by saying that, in order to win, it is sufficient " to be the stronger at a given point and a given moment." He has proved it repeatedly by his leadership in battle.

" Whether we pierce, or simply raise, the veil which, in Napoleon's battles, conceals all the delays necessary for a first orientation or for the arrival of a neighboring

THE BATTLE: DECISIVE ATTACK

force, for some movement, we always find that the decisive attack of the masses is carried out with all its fury and all its tragical characteristics " (Clausewitz).

That is what Napoleon meant also when he wrote to Marshal Gouvion-Saint-Cyr: " One must give preference to no one form of attack, but act in accordance with the circumstances. One must attack the enemy with all possible resources. After having engaged the units nearest to the enemy, one must let them carry on without worrying about their good or bad fortune. But one must be careful not to comply too readily with calls for assistance from the leaders."—" He added," says the Marshal, " that only towards the close of the day, when he noticed that a worn-out enemy had thrown in the greater part of his resources, did he gather up what reserves he might have, in order to thrust on the battlefield a strong mass of infantry, cavalry and artillery; this not having been expected by the enemy, Napoleon caused what he called *an event*, and in this manner he had nearly always been victorious."

Napoleon knows very well that one cannot demoralize or defeat a real opponent by lines of skirmishers or even by a general attack. To overcome him, to compel him to acknowledge defeat, an unexpected effort of unknown violence must be made; Napoleon obtains that effort from the mass, from masses; and to assure the effect, he disposes these masses in columns.

War is like the other human activities: in the presence of new difficulties, of ever greater obstacles, it returns to its origins, to its primary nature, all of violence. It seeks there the means of better surprising in order the better to crush, to destroy the enemy's morale. Along these lines, Napoleon increases continuously the simplicity, the brutality, the vigor of his attacks.

Decisive attack, such is the supreme argument of modern battle, struggle of nations fighting for their existence, their independence or objects less worthy; fighting in any case with all their resources, with all their passions, masses of men and of passions which must therefore be crushed.

If we could study in detail the atttack made by Macdonald's masses at Wagram against Archduke Charles, we should find it:

Prepared by a charge of 40 squadrons (to clear for it a place of assembly), and by the fire of 102 guns (to halt and shake the enemy);

Carried out by 50 battalions (22,500 men). We should find that mass of infantry: Powerless to act by fire, because of the formation it has assumed;

Without effect by the bayonet: nowhere does the enemy await the shock;

Doing absolutely no harm to the enemy, while suffering much itself;

Reduced to 1,500 victorious men when it reaches its objective at Süssenbrunn.

In short, the decimated troops beat the decimating troops, and they decide the army's advance, that is victory on the Marchfeld. The result was obtained not from material effects—they are all in favor of the defeated side—but from a purely moral action.

To this battle of maneuver characterized by one supreme effort, the decisive attack producing surprise, there

THE BATTLE: DECISIVE ATTACK

has often been opposed the battle in line, in which the engagement is general, and in which the commander relies on some favorable circumstance or happy inspiration, which generally do not appear, for the choice of time and place of his action. He may even depend for this on his subordinates, who in turn depend on *their* subordinates, so that finally the battle is won or lost by the rank and file.

History records several successes from the employment of this method. It is not surprising that, particularly in an army like ours where the native qualities of the race cause in all ranks treasures of initiative, of merit, of spontaneity, one should see success result from the natural employment of these qualities rather than from efficient leadership of the higher command.

In every lottery there are fortunate men who win a prize, yet no sensible person depends on lotteries as a means to fortune. Certain causes independent of our will, including chance and happy initiative, sometimes determine events, but they cannot be depended on, and still less be used as the basis for action.

If we analyze this battle in line, what do we find?

The engagement is general, and needs to be supported everywhere; forces being used up, they are renewed, replaced or increased. The result is a constant wearing-down, against which one struggles until the result is obtained from one or more lucky actions of the troops, subaltern leaders or soldiers, always from some source of secondary importance which can only employ a part of the resources available.

The total is made up of a series of more or less similar minor battles, out of the control of the higher command.

It is an inferior form of battle, therefore, if we compare it with the battle of maneuvers, depending on the leadership of the Commander-in-Chief, on judicious and combined use of resources at hand, on the value of *all* these resources, true economy of forces aiming at the concentration of efforts and of masses on one chosen point. Till the last it remains a single combination of combats differing in their intensity, but all aimed in one direction for the purpose of accomplishing one final result: *the foreseen, determined and sudden action of masses employing surprise.*

The weakness of the battle in line lies in the fact that it is an attack which develops everywhere with equal force, resulting in uniform pressure on an enemy who opposes a resistance equally uniform, but of superior value because he disposes of special advantages, cover, fields of fire, etc., which the attacker does not possess to the same extent.

But if we can perceive the weak point in the enemy structure, or a point of little resistance, the equilibrium is broken, the mass rushes through the breach, and the obstacle is carried. If we seek the weak point, or if we create one by our blows on a part of the enemy's line, we attain the same result.

Mechanics and psychology lead us both to the battle of maneuvers. One recommends the application at one point of superior forces, the other urges the apparition of a danger of an attack which cannot be parried. Both therefore mean decisive attack.

Such an attack is necessary, because without it nothing is accomplished, and we can depend only on luck. It is sufficient because it brings the desired result.

THE BATTLE: DECISIVE ATTACK

Theoretically, a battle begun is an attack determined to succeed.

Theoretically also, to be the stronger on a given point at a given time, we must apply all the forces simultaneously on that point, and in an unexpected manner.

When we pass to practice, we shall find that this necessity entails others; the principles of protection will appear again, and compel sacrifices, absorb forces.

To direct the attack, to guard it against the enemy, to prevent that enemy from carrying out a similar maneuver, we shall have to undertake and carry through many minor engagements, each one having some special purpose. Nevertheless, the decisive attack is the keystone of the battle, and all the other combats must only be considered and organized in the measure in which they facilitate and assure the development of the decisive attack characterized by mass, by surprise and by speed, for which we must consequently reserve the greatest possible number of forces and of troops with which to maneuver.

Hence economy of forces, meaning their distribution and employment in battle.

The difference between the battle of maneuvers and the battle in line does not consist merely in the difference of results: results planned and sought in the one case by a decisive attack; results hoped for in the other case from some happy occurrence on one or several unknown points of the front. There is also a complete difference of leadership, of execution, of economy of forces.

That has to be pointed out, because though we theoretically abandon the battle in line, we actually return to it if we have not in advance organized our combination with a plan of battle which aims above all at decisive attack.

In the battle in line, tactics merely consist in overcoming hostile resistance by a slow and progressive wear of the enemy's resources; for that purpose, the fight is kept up everywhere. It must be supported, and such is the use made of the reserves. They become warehouses into which one dips to replace the wear and tear as it occurs. Art consists in still having a reserve when the opponent no longer has one, so as to have the last word in a struggle in which wearing-down is the only argument employed. In that case, the reserves have no place chosen in advance, there must be some everywhere, ready to be employed wherever needed to continue action on the whole front. They are gradually absorbed, and their only purpose is to keep the battle from dying out.

In the battle of maneuvers, the reserve is a sledgehammer planned and carefully preserved to execute the only action from which any decisive result is expected: the final attack. The reserve is meanwhile husbanded with the utmost caution, in order that the tool may be as strong, the blow as violent, as possible.

Finally, it is thrust into the struggle boldly, with a firm determination to carry a chosen point. Employed for that purpose as a mass, in an action surpassing in energy and violence all the other stages of the battle, it has but one objective.

According to Napoleon, there was *no general reserve as such*. He had troops reserved, but for the purpose of maneuvering and of attacking with more energy than the others.

" One often speaks of the use and necessity of strong

reserves. The dogma is closely connected with the theory of progressive consumption of forces; it is considered a sacred dogma. But every reserve represents a dead force. . . . The reserves are useful only on condition of being engaged. . . . One can even imagine a case where it would be wiser to have no reserve; that is where the enemy's force would be precisely known, and when he was already fully deployed " (Von der Goltz).

The difference in employment of the reserves is so great between the two kinds of battle that the other differences are sometimes forgotten.

The battle in line is a principle of the French Army of 1870, or rather the absence of principle as to the conduct of the battle. It is a case of everyone for himself, defeat being always officially due to the arrival of strong reinforcements on the German line; but these reinforcements were precisely troops reserved and brought in numbers to that point to create the demoralization by which armies are destroyed.

This wording of our official reports shows also that if these fresh troops had come to us, it is only as reinforcements that they would have been used, for distribution all along the line, and not as a means to an action of which nobody thought.

XI

THE BATTLE: HISTORICAL EXAMPLE

HAVING seen how theory leads us to conclude that battle is a decisive attack, and that the sole purpose of the battle is to successfully prepare the decisive attack, let us search for examples in history of how battles are planned with the aim of the decisive attack.

We shall, for the time being, take an example on a small scale: the Battle of Saalfeld. We can find there all the actions of battle in reduced proportions.

It is the 9th of October, 1806.

The Grand Army in three columns finishes crossing the Franken-Wald to penetrate into Saxony.

The army's advance guard (1st Army Corps and 3 cavalry divisions under Murat) precedes the center column. It has defeated, at Saalburg and at Schleiz, the Tauenzien Division moving from Hof to Jena.

Of the right column, the leading unit (4th Corps) has reached Plauen; the 6th Corps following, is at Hof.

Of the left column, the 5th (Lannes) Corps, which leads, comes from Cobourg. On October 9th, after a long and difficult march, it has reached Gräffenthal with the Suchet Division and its cavalry; its 2nd Division is 7 kilometers further back; the 7th (Augereau) Corps has reached Cobourg. (*See Map No. 9.*)

Napoleon knows the main enemy army under Brunswick to be between Gotha and Erfurt; its advance guard is at Eisenach.

The army of Hohenlohe is at Jena, with its advance guard at Saalfeld.

He presumes that the enemy generals intend to march on Würtzbourg, the former via Gotha and Meiningen, the latter via Saalfeld and Cobourg.

While continuing to advance his long columns to withdraw them from the mountains and mass them for the purpose of having them all at his disposal, of taking as soon as possible the initiative of attack, he must foresee the possibility of the enemy taking the offensive first. He must guard against attack of the Grand Army's left column by superior forces. Hence the line of conduct which he lays down to Marshal Lannes:

(1) Have himself joined by the 7th Corps, and attack if the enemy has not over 15,000 to 18,000 men;

(2) In the contrary case, that is if the enemy, having assembled superior forces at Saalfeld, attacks, resist to allow time for the Emperor's arrival with 20,000 to 25,000 men;

(3) If the enemy, hastening the attack, allows no time for the arrival of assistance, retire on Gräffenthal.

The Major-General writes to Lannes:

"Nordhalben, October 9th.

"The Emperor will be this evening at Ebendorf, Davout at Lobenstein, Murat at Schleitz.

"The enemy is presumed to intend the defense of Saalfeld; if he be there in superior numbers, do nothing until Marshal Augereau has joined you. During the day, *we shall have news* of the enemy and, if he should have important forces at Saalfeld, the Emperor will march by night with 20,000 or 25,000 men to reach Saalfeld about noon to-morrow, via Saalberg.

"Under the circumstances, if the enemy masses all his forces at Saalfeld, there is nothing else to do but occupy positions at Gräffenthal.

"The enemy cannot risk advancing on you, having such considerable forces on his left flank; if he should, however, do so in very superior numbers, there is no doubt that you should retire, because he would then be seized and attacked on the flank by the forces in the center.

"But if the enemy only has 15,000 to 18,000 men, you must, after carefully studying his position, attack him; Marshal Augereau's Corps will, of course, be with you. The most important thing in the present circumstances is that three times a day you send news of yourself and of the enemy to the Emperor.

"If the enemy retires before you, reach Saalfeld as soon as possible, and hold favorable ground there."

As we see, if the 5th Corps during its march comes across an enemy in greater numbers, it will avoid destruction by defensive or by a retreating fight.

The foremost bodies, or advance guards, in 1806 are therefore prepared to employ tactics of three kinds:

To attack;

To resist;

To withdraw, that is to maneuver while retreating, in accordance with what they see or learn about the enemy with whom they come in contact.

On the same day, Marshal Lannes writes:

"Gräffenthal, October 9, 5 P.M.

"I have just arrived with the Suchet Division and all the cavalry at Gräffenthal. It is 5 P.M. The Gazan Division will bivouac between Judenbach and the village of Gräffenthal. To-morrow, one hour after sunrise, the whole army corps will be two hours distant from here on the Saalfeld road, awaiting from your Majesty orders which I hope to receive during the day or during the night. . . . A terrible day for the troops and artillery, horrible roads, no resources. . . . Impossible for Augereau to be here to-morrow, twelve frightful leagues from Cobourg to Gräffenthal. . . ."

In fact, on the evening of the 9th, we find:

Of the 5th Corps, the cavalry at Gesseldorf; two divisions in bivouacs as shown above;

Of the 7th Corps, the advance guard in front of Cobourg;

The 1st Division at Cobourg;

The 2nd Division south of Cobourg.

On the morning of the 10th, Napoleon has the following message sent to Lannes:

"Ebendorf, October 10, 6 A.M.

"The Emperor approves the dispositions which you have taken. Hasten the arrival of Marshal Augereau, and attack Saalfeld immediately. The Grand-Duke of Berg and Marshal Bernadotte hold Schleitz."

He also writes to Marshal Soult:

"Ebendorf, October 10, 5 A.M.

". . . Marshal Lannes will only reach Saalfeld to-day, unless the enemy be there in considerable numbers.

Thus the days of the 10th and 11th will be wasted. If my junction is carried out . . ."

When these instructions from the Emperor reach the 5th Corps, it has been on the move for several hours; it has as yet only Marshal Lannes' order.

From Gräffenthal to Saalfeld there are 20 kilometers; 16 kilometers to the outer fringe of the wood. The army corps, starting at 5 o'clock and marching 4 kilometers an hour, can therefore have its head at the outer fringe at 9 o'clock.

The road crosses deep gorges, mountains of no great height, but with rugged slopes generally covered with impenetrable forests. The route is mostly downwards.

The column is in the following order:

The advance guard: Treillard Brigade of light cavalry: 9th and 10 Hussars and 21st Chasseurs, each with three squadrons;

One section of horse artillery (2 guns);

One picked battalion (8 companies from the 4 last regiments of the division);

The 17th Light Infantry (2 battalions and 2 companies).

Then comes the main body, consisting of the 34th Regiment with 3 battalions, and the 40th, 64th and 88th with 2 battalions each; also the divisional artillery of 8 guns and 2 mortars.

There is no interval between the advance guard and main body. None is needed. If the advance guard closes on its advance party while the main body closes on its head the commander has a space of 1,500 to 1,800

THE BATTLE: HISTORICAL EXAMPLE

meters for maneuvering, to withdraw or engage his troops under cover from the enemy's guns.

On a fine autumn morning, therefore, before daybreak (5 A.M.) the start has been made at a smart pace.

The troops are well laden: three days' provisions on the sack; and if they only carry three, it is because they have already eaten five days' supplies, out of the eight given them, before starting: four days' biscuits at Würtzbourg, and four days' bread at Schweinfurt.

However, the march is satisfactory. It is the Grand Army in good shape, and the men sing happily.

At the first halt, the Emperor's proclamations are read to the troops: one to the army, one to the inhabitants of Saxony which they are going to traverse. They are received with shouts of " Vive l'Empereur! " and the march is resumed.

At the head of the troops is Marshal Lannes, brilliant advance guard commander if there ever was one, whose calm and caution we shall presently admire, together with his decision and energy. He is just thirty-seven years old.

It is of him that Napoleon, who was a good judge of men, wrote:

"He was wise, cautious, bold before the enemy, of wonderful coolness. He had had little education; nature had done everything for him. He was superior to all the generals of the French Army on the battlefield for maneuvering 25,000 infantrymen. . . ."

His Chief of Staff is the old man of the column: it is General Victor, 40 years old. Then come the Division Commander, Suchet, 34 years, Brigadier Claparède, 32 years, and Brigadier Reille, 31 years.

Between 6 and 7 o'clock, a few distant shots are heard.

In advance of the column, patrols of the light cavalry are moving, searching in every direction. They are detailed and supported by a party of 1½ squadron, following the road of Eiba and the road of Wittzensdorff, Wittmansgereuth and Beulwitz.

All that reconnaissance cavalry has started very early. On the previous day it had already been pushed well ahead.

Seeking to issue from the wood, it has met on the Ausgereuth road some enemy patrols, which have retired on Garnsdorf.

The roads through the wood, right and left, and the wood itself are free, but a few enemy squadrons have been seen above Saalfeld. In the north, a long column is observed moving from Schwarza to Saalfeld.

The cavalry brigade, by a short trot, issues from the wood, followed by the picked battalion which accelerates the pace. The remainder of the column continues its march, careless and gay.

The marshal reaches the outer edge of the wood.

A few of the enemy's outposts are near the road, and on the heights above Saalfeld. They are attacked with infantry, about 9 o'clock, and thrown back. The advance guard arrives opposite Saalfeld about 10 A.M.

Being out of the wood, observation becomes possible. Marshal Lannes, at the head of his infantry, stops on the height above Garnsdorf. From there he sees:

Three kilometers away, the Saale; on the river Saalfeld, a place of 2 to 3 kilometers circumference, 100 meters below the wood; Garnsdorf, halfway up the slope: further north, the small valley of the Siegen-

THE BATTLE: HISTORICAL EXAMPLE 335

bach; another small valley, of the Beulwitz; Beulwitz, Crösten, Wolsdorf. Further north still, the ground rises to form a spur which commands the elbow of the Schwarza and its junction with the Saale; the highest point is the Sandberg. The country provides absolutely no cover.

To this panorama is added a view of the enemy's army. At the foot of the slopes, on three lines very evenly drawn, appear forces which an experienced eye can estimate at 6,000 or 7,000 men. They form the division of Prince Louis of Prussia.

Saalfeld is held by the enemy.

Some squadrons are maneuvering above the city, towards the opening of the Saale.

What is happening on the enemy's side?

Since October 7th, the division of Prince Louis, advance guard to the Hohenlohe Army, formed of 18 squadrons, 12 battalions and 27 guns, was resting north of Rudolstadt. Its outposts extended from Oberhof to Kahlerten, with supports at Appurg (5 squadrons) and at Blankenburg (3 battalions, ½ battery and 3 squadrons).

On the 9th, Prince Louis, learning of the arrival of the Lannes forces at Gräffenthal, assembles his division at Rudolstadt, and occupies Saalfeld with the Blankenburg supports (less 1 battalion) and 12 guns. There is, therefore, at Saalfeld since the evening of the 9th:

2 Prussian battalions;
½ light battery;
1 heavy battery;
1 company dismounted Chasseurs;
3 squadrons Hussars.

On the same day, 9th, Prince Louis was ordered by Prince Hohenlohe to come to Pössnech via Saalfeld, as soon as his posts of Blankenburg and Rudolstadt would have been relieved by detachments from Blücher. Moved by other reasons, he decides to forestall the attack, and to engage battle for the purpose of saving Saalfeld, which contained some stores.

On the morning of the 10th, informed early of the French Corps' advance on Saalfeld, he moves via Schwarza on Saalfeld.

About 9 o'clock he reaches the height of Wolsdorf, while on the heights of Saalfeld occurs the small advance guard engagement which I have mentioned. He forms his division in three lines, on an elevation which has Crösten in front and to the right, Graba behind and to the left.

That is what Marshal Lannes sees as he reaches the plateau.

The Prussian division has its back to the Saale, having for withdrawal purposes in case of retreat only the bridge of Saalfeld or that of Schwarza. Its strength is easily ascertained. It cannot be reinforced for a long time. Lannes will attack, in accordance with the spirit of the instructions received.

What, on the other hand, does Prince Louis intend to do?

With thoroughly Prussian instinct, he has abandoned to the French the difficult slopes which rise towards the woods, and he has sought the plains, keeping the hollow of the valley where maneuvering is easier. For it is a principle in the Prussian army that one must attack; to attack when the enemy issues from a poor position, and

THE BATTLE: HISTORICAL EXAMPLE 337

to attack in echelons, is the latest idea. To carry out maneuvers, space is necessary, and they do not know how to fight otherwise.

Moreover with the eighteenth century ideas prevailing in the Prussian army, no doubt is felt that the French will take Saalfeld for an objective. Saalfeld is a depot, a junction of roads, a passage across the Salle, true geographical objective. While they march on that point, the Prussians can make a flank attack. But unfortunately for Prince Louis, the generals made by the French Revolution ignore the science of military geography, negation of battle. They only know and only wish one thing, to defeat the enemy.

The Prussian army does not only lack straightforward ideas; it is also short of provisions. For instance, in that land of prairies and in October, they can hardly feed the horses of this small division. During the battle, in fact, an order arrived to " distribute with the greatest care " the forage rations which did not exist. Red tape could do no more.

Although, as regards the French, the striking picture of that division on the banks of the Saale has greatly reduced the difficulties of a first reconnaissance, all necessary dispositions have been taken to overcome such difficulties, should they occur:

Protection has been provided to right and left;

If the patrols are insufficient, they can be supported by the cavalry brigade;

The latter has been reinforced by a picked battalion.

Moreover, there is artillery to both sweep the ground and to resist.

It has been necessary to tear away the mask formed

by enemy outposts at the outer edge of the woods; the advance guard took immediate action and, thanks to its composition, succeeded in obtaining a clear view towards Saalfeld and Crösten at least.

Detachments of light cavalry have also occupied Beulwitz and the eastern point of the forest, by the opening of the Saale. They spread out from there, to confirm what is already known of the enemy.

Under these conditions, and Lannes having decided to attack, how will action be developed against an enemy so prettily aligned at the foot of the slopes?

Before organizing the attack, one must first determine its direction. Shall it be by the right? There is no space for maneuvering; a big strong point, Saalfeld, must be carried first.

Shall it be by frontal attack? That is seizing the bull by the horns, it is allowing the enemy to profit by his formation and field of fire. It is attacking him where he is strongest.

By the left? There are found covered approaches and an easy field of maneuver, wide, without obstacles and having natural cover.

In that direction, therefore, can the attack be prepared without being seen; it can be begun without danger of serious obstacles; it can be given all the importance that is possible with the forces available.

It is 10 A.M., the French column arrives in good order, but more slowly in the heat of the day and on congested roads; three or four hours must pass before all the forces are massed on the ground chosen.

But during that long interval, the enemy may attack

THE BATTLE: HISTORICAL EXAMPLE

the column as it debouches; that must be prevented; it is the duty of the advance guard.

To come on guard by seizing every means of halting the enemy's advance, that is then the first step in preparing for battle. Hence the occupation of crests which permit the use of fire; hence the occupation and fortification of the points which increase the powers of resistance of the troops.

The enemy is massed, and may also move, undertake some maneuver, change in some way the dispositions against which an attack is being prepared. How can that be prevented? By attacking him, but without risking anything; with weak forces and along an extended front, to economize forces. Hence the offensive by small units moving from the points still occupied. Lines of skirmishers will move through the gardens, the orchards, the hollow roads, to threaten the enemy and extend the limited action at the outskirts of points occupied.

In such manner, these points, centers of resistance at first, later become the starting points of offensive actions.

In brief, positions occupied which dot the ground with strong points, connected by lines of skirmishers, with good observation from the heights, acting under cover and supplying the elements of partial offensive: such is the front line.

Further back is a reserve of mobile troops. In this case, it will consist of cavalry. Then, when the first needs have been attended to, infantry will be used. In the same way, moreover, must the advance guard reconnaissance be understood.

In the case which we are considering, such reconnaissance is quite unnecessary, inasmuch as the enemy is clearly seen in the plain, his formations and numbers being visible. In the presence of a concealed and covered enemy, it will have been necessary to reconnoiter, that is determine the distribution and importance of his forces, to be able to employ intelligent maneuver.

How would the offensive have been carried out? Evidently by offensive.

Such offensive, to entail no risks, would have started from the strong points occupied. It would have successively developed on the enemy's front. In any case, it would have borne only on that portion of his front which could interest our attack, that is the portion which we intended to strike, and from which we might be struck.

The enemy is therefore reconnoitered and immobilized on the front which concerns the action undertaken. In that way must we interpret Napoleon's saying: "One engages everywhere." On that principle, a division intended for attack will not reconnoiter a front of 6, 7 or 8 kilometers, in which it has no interest. In the same way, when the reconnaissance has provided enough information as to the part of the line intended to be struck, it stops.

These considerations having been made, Marshal Lannes orders the following dispositions:

(1) The picked battalion will continue to thrust back on Saalfeld the Prussian posts which hold the heights, and it will stop at Garnsdorf, occupying it firmly.

The cavalry will place itself in reserve in the hollow of the small Siegenbach valley, and then:

The 17th Light Infantry will occupy with its two picked companies the northeastern point of the wood, opposite

THE BATTLE: HISTORICAL EXAMPLE

the opening of the Saale, which they must defend;

The remainder of its forces will hold Beulwitz, reaching it by the edge of the woods;
The section of artillery will fight at Garnsdorf;
General Victor will command in Garnsdorf and to the south;
General Claparède will command in Beulwitz.

(2) The remainder of the division will move through the wood, or along the edge of the wood, towards Beulwitz (the remainder of the division consists of 4 regiments and the artillery).

(3) The strong points of Garnsdorf and Beulwitz will be connected by skirmishers. Behind these skirmishers will be the cavalry.

Then, when the troops that have arrived are numerous enough, one will consolidate this weak front with the last two battalions to reach the ground.

As we see from all this, the operation aims from the start at realizing the main effort of the whole division in the direction of Beulwitz.

In consequence of these dispositions, the enemy's protective troops soon reported the arrival of French troops from the three directions, causing much surprise to the Prussians who have a single column.

About 11 o'clock, the French had installed:
2 companies at the southern point of the wood;
1 battalion and 2 guns at Garnsdorf;
2 battalions at Beulwitz.

From the point of the wood to Beulwitz, the distance is 3½ kilometers; along that front are spread out 3½ battalions and the cavalry brigade, with rifles whose effective range does not exceed 200 meters. Such dispositions are frequently discussed, however, even to-day when we have rapid-fire arms effectively sweeping the ground at 1,500 meters. These dispositions are criticized in the name of regulations which prescribe for a battalion a front of 300 meters and no more. But those poor regulations were never intended to forbid what Lannes did. Remember that there is no question for the present of defeating the enemy, that no battle front is therefore considered, but that it is a mere matter of seizing some ground and establishing guardians on it to keep out intruders.

We shall soon see the attack begin, and then we find average lengths of front far less than 300 meters.

The situation on the French side will remain unchanged for a long time. Meanwhile *preparation* is going on.

Simonnet's section of artillery, thanks to the favorable position, keeps up its fire against a much stronger artillery (12 guns, 1 heavy and ½ light battery) and against the 2 battalions from Saalfeld which hold in the open field the approaches to that place. The artillery is assisted by a line of skirmishers on the crest, and by the occupation of Garnsdorf. Still better, it can advance shortly after.

In Beulwitz, the 17th Regiment has occupied the village and thrown out skirmishers who make use of the gardens and orchards to reach the approaches of Crösten. The regiment is covered to the north by cavalry patrols.

At Beulwitz as at Garnsdorf, snipers who have crawled forward keep up against the enemy, unprotected in the open, a murderous fire.

It is the battle for supremacy of fire. In consequence

of the advantage of preparation after drawing up plans, the program can safely be developed, and 10 o'clock finds the majority of French troops duly arrived as per orders.

As regards the enemy, what impression does he receive from all these occurrences? He will answer that question himself.

All information concerning him is taken from the works of an actual witness, the Saxon engineer Mümpfling, who wrote the *Vertraute Briefe*.

After describing unfavorably the situation of the Prussian army, he adds:

" Do you see us later spread out along that threatening rampart, in the open along the thin border of meadows which separates it from the Saale, against which we have backed? From that rampart the enemy snipers, perfectly covered, picked us out at leisure, without possibility on our part of replying to quite invisible people, and this amusement lasted *several hours*. Meanwhile, the French leaders, perfectly located to discover the weak points in our line, made plans in consequence. . . .

" The maneuvers of the Frenchmen were developing more and more; their purpose was to keep at a distance the troops on the left flank, above and near Saalfeld, and to occupy with their skirmishers the whole front of the Prince's position, in order to envelop him and cut him off from the Schwarza. . . ."

This is written, of course, after the event; however, the author has understood the purpose of all the preparation. But during the action, the plans of the French were not so evident. Besides the invisible snipers whose bullets are received with no possibility of returning them, nothing is seen on the horizon; the reconnaissances sent have merely perceived, early in the day, some columns marching towards Eiba, on the main road, towards Beulwitz.

These columns have all disappeared, and only insignificant attacks occur. What are then the intentions and movements of these many columns, of which nothing is seen? In case of a defeat, can the Saale or the Schwarza be safely crossed? Evidently not. A battalion (2nd Müfling) is sent to Schwarza.

At the same time, the order arrives from Prince Hohenlohe to remain at Rudolstadt and refrain from attacking, as the army is to move from Blankenhayn to the Saale. The retreat through Schwarza, in case of defeat, becomes still more important. The Prince occupies the Sandberg with the dismounted battery and the 1st Battalion Müfling.

The Prince-Clement Regiment will place one battalion (the 1st) between Aue and Crösten, to connect this occupation of the Sandberg with the main body of the division. The 2nd Clement Battalion will climb the Sandberg, placing itself on the right of the battery and of the 1st Müfling.

Such is the kind of dispersion inevitable when tactics depend on considerations of ground, become of capital importance.

Instead of a military idea whose realization is sought, obtaining from the ground the material facilities to that end, it is here the ground which dictates the line of conduct. Thus are points occupied first because of their intrinsic value, without measuring their importance in relation to any operation; then, holes are filled up in this occupation of ground; finally, impotence through dis-

THE BATTLE: HISTORICAL EXAMPLE 345

persion of the forces results for the time when action is undertaken.

In this case, 2 regiments and 15 guns are devoted to the occupation of Schwarza and of the heights which command it; 2 battalions and 12 guns are similarly employed for Saalfeld.

It is about 1 o'clock. After making all these concessions to the value of positions, Prince Louis, a bold man much worried by complete doubt, decides to attack. He attacks straight ahead with everything he has at hand: 6 battalions out of 12 (without artillery or preparation of any kind); 4 are in the first line, 2 in second line.

Hardly is this attack perceived, proceeding east of the Crösten-Beulwitz line, before the right comes under fire of the French skirmishers who fill the gardens, the orchards, the hollow roads, ever invisible skirmishers who deluge the right with bullets. The line hesitates, stops, answers by volleys, without any result; at the same instant its flank is attacked by 2 battalions of the 34th which, after proceeding under cover of the slopes, appear and charge in column, with beating drums, General Suchet leading.

The strain is too great. The Xavier Regiment is in full retreat, and the left of the line (Regiment of the Elector) also withdraws. The 17th, on their heels, throws itself into Crösten, where it is attacked in turn by the Prussians reorganized, particularly by the Elector's Regiment which has not suffered and now attacks in flank. Its ammunition giving out, the 17th loses Crösten, and retires on Beulwitz, where it is relieved by the 64th; it passes into the reserve.

Marshal Lannes then sees the situation ripen.

It is approximately 2 o'clock:

(1) All the troops are on the spot;
(2) The enemy's force and positions are known;
(3) He is immobilized;
(4) His forces are dispersed, and already shaken.

The Marshal will attack in the direction already chosen: by the region of Beulwitz and of Crösten. He will strike the enemy mass in the plain with the main body of the division's forces, the Claparède Brigade (17th and 64th) attacking on the front, and the Reille Brigade (34th and 40th) attacking on the flank.

But he must first guard against the troops holding the Sandberg and Aue, and give to the division's attack sufficient space for its development.

The Reille Brigade is intrusted with this double mission. For that purpose, it starts in the direction of the Sandberg, protected by many skirmishers towards the wood of Aue, the 34th in the first line. The 40th is in echelons behind and to the left; the 21st Chasseurs in echelons behind and to the right.

The attack first strikes the Clement Regiment, then the Sandberg battery of 15 guns is seized; it insures the possession of Aue and Sandberg; it carries on the pursuit with part of its forces, and resumes with what remains the flank attack with which it is intrusted.

The time has come to end it all. It is nearly 3 o'clock. The maneuver will at last be carried out, to the construction of which work has been going on since morning.

The artillery has meanwhile arrived; it takes up positions near Beulwitz, then advances and prepares by its fire the attack of the infantry.

Marshal Lannes orders the charge sounded all along

THE BATTLE: HISTORICAL EXAMPLE 347

the line, and one sees then against this shelled opponent " masses of infantry arrive, which, descending the heights rapidly, rush like a strong torrent on the Prussian battalions, and smother them in an instant " (Marbot).

Which is also clearly pictured by the engineer Mümpfling when he writes:

" About 3 o'clock, the French columns rushed on us like an avalanche. Within a second we were cut into three pieces, surrounded with a circle of fire and backed up against the river."

At the signal of attack given by Lannes, all the troops near Beulwitz have advanced, pushing forward:

A frontal attack, composed of:

On the right, the 9th and 10th Hussars;

In the center, the 54th Infantry;

On the left, the 21st Chasseurs;

In the second line, the 87th Infantry and 17th Light Infantry;

A flank attack: 34th Infantry (3 battalions).

It is *all* the cavalry, *all* the artillery less two guns, and four infantry regiments out of five that attack an already shaken enemy. They attack by surprise, that is with an undeniable superiority of resources, suddenly and at close range, that point of the enemy's line which has been chosen as the easiest to assault and specially prepared for the attack. As to frontage, the attack has 1,500 to 1,800 meters for all the troops taking part: that is less than the 300 meters per battalion of the regulations.

It is particularly the French left that strikes; it is most advanced. On the right is some cavalry supported by the infantry; it has descended in mass into the valley.

This cavalry soon seizes a favorable opportunity of

charging the Prussian infantry, which is hard pushed on every side by the French infantry and shelled by artillery. It charges, and sabers right and left for half an hour. Prince Louis of Prussia, seeing his infantry beaten, runs to his squadrons near Wolsdorf, and rushes on at their head, but in vain. He runs across the 10th Hussars and is hard pressed by a Frenchman who calls out to him to surrender; he replies by a cut of his sword, and falls pierced in his turn.

The defeat is complete.

There remains nothing but fleeing men, escaping as best they can towards Blankenberg, Schwarza or across the Saale.

At the time of the beginning of the general attack, Victor in Garnsdorf has assembled his picked battalion, the two companies of the 17th, and advanced on Saalfeld which he has carried; he continues the pursuit as far as Rudolstadt along the right bank of the Saale.

The pursuit has been continued also towards the Schwarza. Claparède leads a whole brigade (17th and 34th) there, pushing back the enemy past Blankenberg, and crossing the Schwarza by wading to the waist. Three captains of the 17th are mortally wounded there.

The day's trophies include 1,500 prisoners, 4 flags, 25 guns, 2 mortars and 6 gun-carriages.

The Prussian losses in killed are not exactly known, but Lannes, writing that same evening to the Emperor to report the battle adds, though he had no sensitive soul: *The battlefield is a horror.*

The Suchet Division, which has been the only one engaged, has 172 casualties and 10 horses killed.

Simonnet's section of artillery has fired 264 shots. The divisional artillery has fired a little less, about 236 shots. The infantry has used up about 200,000 cartridges, an average of 20 per man.

Such is the result of the wonderful methods employed by the young marshal. One wonders what to admire most, of that wisdom which patiently prepares the battle during six hours, or of the *à-propos* and energy with which he starts his final attack. So true is it that the art of fighting, even for the most ardent and most energetic leaders disposing of the best troops, does not merely consist in rushing at the enemy in any sort of way.

The theory adopted is evident in this case; we clearly see the lengthy maneuver (from 9 o'clock to 3 o'clock), aimed solely at the overpowering ending by the main body, preceded by a preparation for which the *least possible forces* are employed.

The Battle of Saalfeld, if delivered to-day, would entail no different method.

The advance guard would seize Garnsdorf and Beulwitz, and protect itself towards the Saale.

It would be reinforced by all, or part of, the artillery, according to necessity, and it would assume either the offensive or defensive against the enemy, as necessary:

(1) According to whether he is covered from fire, covered from sight or reconnoitered;

(2) According to whether he attacks, or merely resists;

(3) According to whether he maneuvers, or stays still;

(4) According to whether he engages his forces or economizes them.

Under protection of that combat of advance guard, and then of the whole front, the main body is brought to the point where one wishes to produce the principal effort.

That point would evidently depend on the same considerations. The direction chosen for attack must offer:
(1) Good approaches;
(2) Few obstacles;
(3) Space to maneuver.

Of the main body forming the reserves one would make two parts: the principal one assigned to the decisive attack and to the measures of protection which it implies, a weaker one destined to keep up the line of frontal attack.

This idea of decisive attack had been thoroughly learnt by the Germans of 1813 from the study of the wars of the Empire, as shown by the Instructions to Officers Commanding Corps, Brigades, etc., issued by King Frederick-Wilhelm during the truce of 1813:

"As I have noticed, in battles, that the various arms have not always been suitably employed, and that the dispositions for battle are usually unsatisfactory, I wish, in connection with the early resumption of hostilities, to repeat the following *rules of war*.

"They are general principles:

"(1) According to the manner in which our enemy makes war, it is usually inappropriate to open the combat with cavalry, or to bring immediately all troops into action. From the way he employs his infantry, he succeeds in delaying and strengthening action; he seizes villages and woods, hides behind houses, hedges or ditches;

THE BATTLE: HISTORICAL EXAMPLE 351

he knows how to defend himself skillfully against our attacks by his own attack; he inflicts losses on us with few troops when we advance against him with strong masses; he then relieves these troops, or sends fresh troops into the fight, and, if we have not on our side fresh troops to oppose to him, he pushes us back. We must therefore adopt this principle, which is a principle of the enemy, to preserve our forces, and to feed the battle until we pass to the *main attack*.

"(2) Our artillery has had little effect, because we have divided it too much. . . .

"(6) War in general, and especially the result of battles, depends on the superiority of forces at one point.

"(7) To obtain such superiority of forces, it is necessary to deceive the enemy as to the point of attack, and to make both a feint attack and a true attack.

"(8) Both attacks must be concealed behind skirmishers, so that the enemy may not distinguish between them.

"(9) A line of skirmishers is used first. The enemy's attention is drawn by several battalions skirmishing on one of the flanks, which is vigorously shelled at the same time.

"(10) Meanwhile, the true attack is retarded, and it only begins later, when the enemy's attention is entirely engrossed by the feint attack.

"(11) The true attack is carried out as rapidly and as vigorously as possible, by a great mass of infantry, superior in numbers, when possible, while another body turns the enemy's flank. . . .

" In a general way, one assigns:
" 1 Brigade to the feint attack;
" 2 Brigades to the true one;

" 1 Brigade in reserve.

" These principles are known to you, and have been recommended to you several times. We have often practiced them in our peace maneuvers, but I remind you of them as things known are sometimes forgotten, because by their simplicity they seem evident, yet victory often depends on them. Unless care is taken and the memory refreshed daily, one tends to over-wise combinations, or else one marches into battle with no combination which is worse."

As we see, after developing the theory of battle of preparation, or feint attack, and of the decisive attack which he names *true attack,* after having explained how it is fought, the King points out how the forces can be subdivided. Later, this is changed into the formula, subject to modifications of course, of:

1-3 to engage;

1-3 to wear out;

1-3 to finish up.

If we go back to the 18th of August, 1870, we find this idea of a decisive attack to be prepared as basis of all Von Moltke's combinations.

XII

MODERN BATTLE

Execution

IF, from the Battle of Saalfeld previously studied, we pass to the use of modern armies, a number of factors are modified:

(1) Weapons have a greater range; they are more deadly, and that compels taking the dispositions for attack further and under better cover. In the same way, during the action, one must seek more than in the past to make full use of the power of weapons.

(2) Armies maneuver easily, and are carefully protected. The enemy's dispositions are therefore harder to ascertain; reconnaissance must be prolonged; immobilizing the enemy is also more difficult.

(3) The effectives on either side reach unknown proportions.

It would take too long, therefore, to mass the troops for a decisive attack. These troops, once massed, can generally be employed only on the ground they occupy, for lack of time to move them elsewhere. Hence the need of knowing a long time in advance the exact direction in which the attack will be carried out, and the need of a more thorough reconnaissance of the enemy and of the ground.

In any case, there is a great increase in the preparation, which must

        Inform   better;
        Resist   longer;
        Immobilize   more.

Moreover, the conduct of the attack, once the direction is determined, will necessitate finer tactics, aimed, besides moral effect, at the complete use of perfected material resources available.

Nevertheless, all that has been said by me as to the theory of battle and the means it employs remains fundamentally true, inasmuch as the same moral creature, man, still wages it. The various phases of the battle therefore remain the same:

    To prepare ⎫
    To execute ⎬ the decisive attack.
    To profit by ⎭

The duty of the command consists, as in the past, in planning these phases, and in adopting a distribution of forces aimed at:

(1) Obtaining protection from the enemy, opposing to him for that purpose, at every point where he appears, forces capable of resisting as long as the preparation lasts;

(2) Organizing the decisive phase, while holding back a mass ready for contingencies or for intervening at the opportune time, whether to parry or to strike.

That distribution, in order to comply with the principle of economy of forces, allots:

To preparation, the least possible number;

To the execution, the most forces possible;

To the exploitation all that are still valid, without possibility of laying down any absolute rules or mathematical equations of such distribution, especially as regards the final reserve; its strength depends, of course, on the information then available as to the enemy, on circumstances and on the temperament of the commander.

Remember Bonaparte engaging the very last soldier at Aboukir, and Napoleon in 1812 saving his reserve at the Moskowa, because instead of being a young and ambitious general with nothing to lose and all to gain, he has become an Emperor with much to lose.

We will take up again, under the new conditions, the study of the preparation and execution of the decisive attack, to find out what tactics are needed for these two phases of battle, and what use should be made of the various arms.

Preparation

By *preparation* are meant all the dispositions aimed at directing and insuring the execution of that decisive phase which is the true reason of the battle. It is a development of the idea of protection already familiar, taken in its widest sense with the object of acting on reliable information, and maneuvering without danger of the enemy's blows as long as one is in no position to return them with interest.

The main object, therefore, is to supply to the commander such information as he needs to direct and execute with full understanding the decisive part of the battle. Considered from this point of view, it entails the search for the objective to strike, for the ways and means of reaching it, as also for knowledge of the enemy's condition. It represents a mission of direction and information which continues till the moment of the decisive action, and which often begins long before the battle.

In like manner, the strategic advance guards of Napoleon, especially those of 1806 and 1809, supply by their information the basis for Napoleon's maneuvers, as they later, by their resistance, constitute the pivot around which the maneuver is developed.

In like manner also, is the battle organized which Von Moltke sought on the Sarre about the 9th of August, 1870. He relies on the information he has concerning the French Army to attack its front with his own 1st and 2nd armies, while the 3rd, debouching from the Vosges, is to carry out a flank attack which will be the decisive phase of his combination. Preparation is begun in that case long before August 9th.

At first, preparation is interwoven, as we see, with information resulting from exploration; it therefore is a duty of the cavalry, assisted by the artillery.

As the enemy draws nearer, information becomes a task of the protective troops. But as preparation must also assure to the commander his freedom of action until the last minute, allowing him to make up his mind only when fully informed, preparation and protection are well intrusted to the same hands. The advance guards then interfere to halt the enemy, or at least impede him.

At Saalfeld, for that purpose:

The picked battalion occupies Garnsdorf;

The 17th Regiment occupies the southeastern edge of the woods and Beulwitz.

As we have seen also, Marshal Lannes, if facing a division better concealed, would have extended his reconnaissance; the advance guard would have been more active; it would have reconnoitered towards Saalfeld, towards Crösten, employed its guns against one or other of these places to compel the hostile artillery to show itself. It would have assured, in any case, the occupation of strong points prepared for resistance.

Besides what has been pointed out, the preparation must *conceal* the direction and time of the decisive attack; it must cover its organization, which entails a new duty of protection.

It must also hold the enemy to the situation previously observed, and forbid him to prepare some maneuver of his own. For that purpose it must immobilize him, making it materially impossible for him to assemble enough forces to successfully resist the decisive attack; it must therefore engage the enemy.

Methods of Preparation

For the fulfillment of its double duty, the preparation must attack the enemy wherever he shows himself, with the object of causing him serious losses, of depriving him of his means of action, of so threatening and paralyzing him as to prevent his moving any forces elsewhere. Its attitude must therefore be undoubtedly offensive.

But it must also hold the enemy if he threatens. While acting, it must prepare facilities for successful defense.

To conquer and to hold with increasing energy, such is its mission.

The small proportion of troops at the disposal of the preparation over a generally extended front does not permit of even action along the whole line. The offensive is liable, for that reason, to become distributed; its aim is to win those points whose natural strength or commanding position assure easy conquest and future disposal of the spaces in between.

If, at Saalfeld, the enemy had been efficiently protected and organized, he would have occupied Garnsdorf and Beulwitz. Garnsdorf first, Beulwitz later, would have been attacked, and possession of the interval would have resulted. All available troops would have joined hands, if necessary, to carry the former of these points. That being done, they would have held it with the smallest possible forces, and would have reformed in order to carry the second point.

On a modern field of battle, where resistance is generally planned in depth, behind a first line of obstacles will usually be found a second one. Then the assaulting troops, extended and mixed with one another, must be constantly taken in hand again by the commanders of small units for the purpose of making new efforts against the many objectives which they must successively attack, in simultaneous time, if they are to succeed.

Thus, in execution, the troops of preparation appear to us as engaged not in one action, but in several partial combats, independent one of the others, seeking to overcome the centers of hostile resistance.

As the enemy, meanwhile, observes the same methods until he has been completely immobilized, or as he seeks to win back the points he has lost, there ensues a series of offensive and defensive actions for the possession of ground. This generally gives to the battle of preparation particular characteristics of perseverance, which result in wearing down the enemy.

Hence also the length of the battle of preparation, existing as a constant offensive, everywhere carried on, under difficulties; and in case of a reverse, it becomes a defensive prepared in advance and fought bitterly.

As we have seen, the preparation consists of a number of partial engagements, each of which, in order to succeed, to obtain a decision, entails some decisive action. Every such action is, in turn, divided into Preparation, Execution and Exploitation. At each stage, the use and the formations of the troops are governed by the same principles as prevail in battle.

It is certain also that these numerous engagements cannot all be under one leader.

The higher command then exercises its action by distributing the work of preparation among a certain number of subordinates, leaving to the initiative of each the task of overcoming the enemy by the means at his disposal. It keeps for itself the main task, that of directing and executing the decisive attack, and it carefully keeps the possibility of intervening till the last moment by means of the general reserves.

Finally, preparation results in a general engagement on the whole front, in a struggle often very painful and frequently very lengthy. So that, while theoretically this operation must absorb a minimum of forces, it actually necessitates important sacrifices. The commander must know how to make such sacrifices without stint, as long as they do not compromise the following operations, and particularly the decisive action.

Duties of the Various Arms

Artillery.—It is artillery which can, of course, act first because of its range, its mobility, the ease with which it goes into, or ceases, action in order to move elsewhere when necessary.

For that reason, the main body's artillery, moving for the greater part near the head of the column, increases its speed. Under protection of the infantry, it reinforces the artillery of the advance guard.

To assist the advance guard in its mission of reconnoitering, immobilizing and wearing out the enemy, the artillery must destroy the obstacles which impede the infantry: strong points and hostile artillery.

As soon as progress is possible, it advances in its turn, to settle finally the fate of the enemy's guns by a battle at short range. That is an artillery duel. It is evidently most important to obtain as soon as possible the superiority in such a duel of guns, capable of spraying with fire all the ground within range.

For that purpose, our artillery must at once have the advantage of numbers, organize a long line of fire, bring up all its guns and keep nothing in reserve. Such is the method followed by artilleries fighting an artillery duel.

Once the enemy's guns have been destroyed or silenced, we must resume the function of assistance to the infantry, by preparing the attack of such points as the latter chooses for objectives.

That preparation means, as we shall see later when studying the decisive attack, the opening up of approaches to the objective, and the partial destruction of that objective, to accompany the assault.

MODERN BATTLE 361

This team play between artillery duels and infantry attacks causes variations in the dispositions of batteries. In the former case (artillery duels), the divisional artilleries must strive to join the corps artillery. All the batteries must, as much as possible, fight as one unit, the army corps' artillery, aiming all at one purpose.

In the latter case (the infantry attack assisted by artillery), the divisional artilleries remain, of course, under the orders of the division commanders; they are, moreover, reinforced by part or all of the corps artillery.

As firearms are improved, the infantry is compelled, in order to advance, to travel under cover, at least from enemy artillery. To that end it takes advantage of every favorable means of approach for as long a time as possible. *The necessity of cover is increasing daily.*

But these means of approach are easily held to-day by small units, occupying strong points and armed with repeating rifles, machine-guns, or a few quick-firing guns for enfilading purposes. In former times, many guns were needed for such a result, while now a few are sufficient. Hence the covered approaches so necessary to infantry would be useless if the infantry were not closely supported by an artillery capable of destroying the enemy's means of resistance. The union of both arms becomes more necessary, and only behind a curtain of shells that destroy obstacles and silence enemy guns will the infantry be able to advance.

While this will sometimes necessitate dividing the artillery up into groups having different missions to carry out, we must not forget that the moral effect, peculiar to artillery, increases greatly with the concentration of fire. It is, therefore, by mass action that important decisive results must be sought.

On the other hand, artillery possesses to a very high degree the facility for surprising: it can, as soon as it appears, follow up the threat with the blow, the apparition of danger with fire. It must keep for its action these characteristics, increase them if possible, and for that purpose destroy the enemy as rapidly as possible by intensity of fire.

Infantry.—Although artillery opens the fight, it cannot do so without assured protection. The infantry must, therefore, open up the battlefield for artillery, and protect it unceasingly by holding the necessary points at sufficient distance to guard the batteries against the hostile infantry's fire.

Apart from the duty of protection, infantry has the leading part in the preparation destined to wear out the enemy. That causes special development of the battle for supremacy of fire.

The troops of preparation must also immobilize the opponent, which compels the infantry to attack him, to threaten him with assault after approaching within assaulting distance.

We shall deal with this double action by fire and by advance, seeking how it can be developed, with forces reduced to the smallest necessary numbers, so as to attain the purpose sought: the threat at close range, which is not a decision, yet is difficult to carry out.

It is evident, to begin with, that, to-day, fire direction and fire control have an immense importance. Fire is the supreme argument. The most ardent troops, those whose

MODERN BATTLE 363

morale has been most excited, will always wish to seize ground by successive rushes. But they will encounter great difficulties, and suffer heavy casualties, whenever their partial offensive has not been prepared by effective fire. They will be thrown back on their starting point, with still heavier losses. The superiority of fire, and therefore superiority of control and direction, becomes the most important element of an infantry's fighting value.

No struggle will be possible between a body having no theory, no instruction, no discipline in the use of fire, of little value therefore in action, and another body perfectly trained, disciplined in the use of fire, far superior on the battlefield.

Training and instruction will combine with the nature of the armament in determining the points to be held. But under no circumstances must these considerations limit the offensive, which fire must, on the contrary, assist and develop. Fire is never an end, but only a means of maintaining an advance; it is used for that purpose, and it must be ceased as soon as the result sought—possibility of resuming the march—is attained.

Precede, therefore, any forward rush by a hail of shot destined to shake the enemy and keep him down. It is then over a ground covered with strong points and obstacles of every kind that the advance must be made. For that purpose, preparation must expel the enemy from his cover, seizing and holding the strong points. If they are occupied, they must be attacked by combined action of the units nearest to them, and fortified as soon as they are taken. The troops must then be reformed, and a new start made from the points carried, to seek a further advance against new objectives. A series of similar actions is carried out in the preparation until the last enemy lines are reached, or until the troops can go no farther.

These efforts are mainly the duty of the front line troops, owing to the necessity of keeping the others back, reserving them to feed and maintain the preparation.

They cannot be directed by the higher command, nor by any command acting in depth. Their success, however, requires sufficient forces, so that commanders of front line units (battalions or companies) must show initiative and good team work to combine the action of their respective forces against the successive objectives to be reached.

To be wisely managed, this combat must grow in vigor and energy from its start till the decisive action, which necessitates successive reinforcements, and therefore the disposition of the troops in depth. This is also made unavoidable by the need of preserving the reserves for a long time from the fire which sweeps the first line.

Although this phase of the battle is often considered as a mere demonstration, it requires extreme energy from those engaged in it. For the troops in the firing line there is only one possible attitude, fighting with the utmost determination, with all the resources at their disposal, employing fire, marching, intrenching tools when available.

Every attack undertaken must be carried through, every defense begun must be continued with the greatest energy. Troops only stop when exhausted, or when insufficiently

MODERN BATTLE

supported by the units in the rear, at the disposal of the commander.

Finally, as we have seen, the front line troops have to employ improvised fortifications to protect their gains against the enemy's counter-attacks. Not only will front line units strengthen as best they can the positions they occupy, but those in the second line must also fortify such positions as they reach during the progress of the action. The reserves also, assisted perhaps by the engineers, should organize further lines of defense for use in case of a retirement.

Decisive Attack

The preparation, by its ceaseless offensive, has finally succeeded in:

Thrusting back the opponent's first line;

Carrying his outposts;

Immobilizing him by its efforts and by the threat of an early attack.

But it is exhausted; the greater part of its supports are engaged, units have become mixed, casualties have occurred, and ammunition is running low.

It is facing the enemy's main body, serious obstacles, a ground swept by fire and dotted with strong points.

Before it is a zone almost impassable; there remain no covered approaches; a rain of lead beats the ground. Yet success is not achieved; "nothing is accomplished so long as there remains something to be done." Victory hangs on the enemy's bayonets, and it must be seized there, conquered by man-to-man struggle if you will.

To reinforce the troops of the preparation in order to

get that result would be useless: that would be renewing the useless battle in line.

To flee or to charge, that is all that remains. To charge, but charge in numbers as one mass, therein lies safety. For numbers, if we know how to employ them, allow us, by the superiority of material means placed at our disposal, to overcome the enemy's fire. With more guns, we can reduce his to silence, and the same is true of rifles and of bayonets, if we know how to use them all.

Numbers give us moral superiority by the sentiment of strength which they create, and which we will increase by formations.

Numbers mean surprise for the enemy, the assurance that he cannot resist caused by a sudden apparition of danger, and by the rapidity and proportions of an attack which he has neither the time nor the means to parry.

Preparation by Artillery

"Whoever can move a mass of guns by surprise to a given point is sure of success" (Napoleon).

To make a breach on the front, so as to open a way for the infantry, to keep it open, to sacrifice itself if necessary in order that the infantry may carry out its task, to keep watch on enemy batteries and counter-attacks: such is the artillery's present duty.

For that purpose, the greatest possible number of batteries enter into action on the point selected for attack. There are never too many of them, there are never even enough.

All artillery groups near that point work to the same

MODERN BATTLE 367

purpose, by a sudden and violent fire in constantly increasing proportions.

Fire must be opened up against the obstacles which may delay the infantry's advance, and the first of these is enemy artillery. It becomes the *first* objective assigned to the massed artillery.

Once superiority is obtained in this duel, the obstacles and strong points along the road must be destroyed, or at least rendered untenable.

The road being open, it has to be kept open; for that purpose, fire is kept up against that part of the enemy front which has been chosen for attack until it is reached by the assaulting infantry.

Success must also be assured by shelling any troops which the enemy might wish to use: new batteries, reserves or counter-attacks.

To fulfill this third task, the massed artillery organizes battery groups as attack groups or counter-attack groups, their function being to accompany and assist the infantry columns, and to maneuver in the dangerous directions. As their movements must be very rapidly carried out, it is the duty of the commanders to foresee such movements, and to find positions suitable for each case.

Such unity of action will only be possible for many batteries, of course, through a sole source of command, able to direct and control their fire as well as distributing among them the various duties, and co-ordinating their work with the needs of the infantry attack.

EXECUTION, INFANTRY

During the general preparation of the battle, which has lasted perhaps a considerable time, the infantry troops destined to execute the decisive attack have assembled opposite the objective, employing for that purpose covered approaches. They are massed behind such cover as is nearest to the enemy, protected by the troops of the preparation. They are, of course, covered in every dangerous direction.

When the time comes to act, artillery shakes the enemy's resistance, and infantry must destroy it. To compel the enemy's retreat one must *march on him;* to conquer the position, to take his place, one must *go there.* The most powerful of fires does not give that result, and here begins the work of the masses of infantry. They march straight to the objective, increasing their speed as they draw nearer, preceded by violent fire, in order to assault the enemy bodily and close the argument with cold steel by greater courage and determination.

What formation should be taken depends, of course, on the ground and on the distance at which the infantry appears in the open. It is determined by the general considerations which govern the infantry battle.

First the mass must be moved up to a distance where it can effectively use its weapons. Until that distance is reached, it is powerless to defend itself by fire, but it must suffer some, especially from the artillery whose rapid fire would suffice to its disorganization.

At first then, we see a period during which the mass of infantry, practically unarmed, endeavors by its formation and speed to avoid as much as possible the enemy's fire. Hence combinations aimed at the least exposure and most movement: thin lines, extended order, irregular lines of sections or platoons, etc.

However, in the presence of an enemy with full freedom of fire, formation, even of the most clever kind, will usually not suffice to advance over unprotected ground, or even to cross spaces of any width. Casualties will occur which break up the organization and especially the morale of the infantry.

In the second stage, on the other hand, when close to the enemy's position, the mass is able to develop all its power of fire and of shock. The formations to be employed must seek to utilize, as much as possible, these two means of action, prolonging them without pause, so that their effects overlap.

The fire received has then but secondary importance; a start has been made and the goal must be reached. Moreover, there is but one way of lightening the effects of enemy fire, and that is by developing a more violent fire, capable of keeping down the opponent, at the same time advancing rapidly.

To advance, and advance rapidly, preceded by a hail of shot; as the enemy is neared, to display troops ever more numerous and denser, such is the principle of the formations to take and of the conduct to adopt.

An attack organized as we have seen can only act straight to its front. Abandoned to itself, it has its flanks exposed to the attacks or counter-attacks of the enemy. Yet its only chance of success is a straight, rapid and continued advance. Hence the need to protect it against all surprise, to organize in the rear another body, independent of the attack but capable of foreseeing the counter-attack and preventing it. That is the first duty of the reserve.

Moreover, when the attack has carried the position,

enemy reserves will seek to appear before the attacking troops can have reorganized. These reserves must be fought immediately, especially if counter-attacking. That also necessitates maintaining until the end an important reserve, at sufficient distance to neither be engaged with the attacking troops nor suffer from the same fire.

Cavalry, Exploitation

Simultaneously with the infantry attack, when in the open, there suddenly appear, on the enemy's front, flank or rear, the squadrons of attack. They throw themselves against whatever resistance is still displayed by the enemy, or against any hostile cavalry seeking to charge the assaulting infantry, or against enemy reserves as they hasten up. These reserves alone can change the course of events, and for that reason they hasten. They must be stopped at any price.

Whether the cavalry aims, therefore, at the enemy's threatening squadrons, or at the main position, or at the hostile reserves, it can always find, at a time when the strain is greatest, an opportunity to do some good work.

For cavalry as for the other arms, there is then a need and possibility of action towards assisting the decisive attack, a victory for all resulting sometimes from even apparently unsuccessful efforts of a few.

Choice of the Point of Attack

The decisive attack is the employment of the mass, bringing moral and material superiority.

Moral superiority, resulting from numbers, formations,

etc., is no longer sufficient to-day with modern weapons: their effects are too demoralizing.

We must also develop material superiority, employing to good purpose the many guns and small arms supplied by the mass; this requires space.

Hence the preference, in modern tactics, for flank attacks which permit of developing against one chosen point the desired superiority of fire; of carrying out flank and reverse fires of undeniable moral effect; allowing by an unlimited space the possibility of always maneuvering the mass.

Hence the abandonment of the central attack so often practiced by the Emperor. He who employs it now is enveloped by enemy fire and cannot use all of his own fire.

Yet the nature of the ground has a part in determining the choice of the objective for decisive attack.

We have seen that, until it is close to the enemy, the attack suffers much from fire while inflicting little harm; art consists therefore in decreasing that danger zone, in delivering the attack from as short a distance as possible. Moreover, once the assault has started, it must advance rapidly; that requires ground without obstacles, which does not mean without cover. The ideal is open and undulating ground. The main thing is to waste no time.

Ground may, consequently, determine the point of attack, for if these two possibilities are realized: to start from close range, and to advance rapidly, the dangers of central attack disappear.

It will also be often necessary to employ the masses where they happen to be, whether their presence be due to more or less exact knowledge of the enemy's situation, or to location of ways of communication, or some other cause. They cannot be rapidly moved from left to right on the battlefield because the extent of the fronts does not allow it any more.

Finally, decisive attack in the battle of to-day is not to be sought indifferently on any point, yet it can be imposed on us, or an opportunity suddenly arise, in spite of all that theory may teach.